"This important book offers a comprehensive examination of the link between environmental conservation and wellbeing. It delves into how eco-friendly practices improve both community ties and individual mental health. The authors show that caring for the environment and biodiversity enriches our lives with purpose and happiness, utilising theories and practical examples from positive psychology. It illuminates the links between environment, psychology and wellbeing in an accessible way and will be a valuable resource for academics, students and practitioners."

Professor Conor Murphy, *Department of Geography, Maynooth University, Ireland*

"An essential read for anyone invested in the intricate dynamics of water systems, this book offers a compelling case for holistic understanding and community-driven action. Through insightful analysis and captivating examples like the Let it Bee project, the authors underscore the urgent need to view rivers and their surroundings through a broader, systems perspective – one that encompasses environmental health and human well-being. Whether you're a student, practitioner, policy maker, or scientist, this interdisciplinary gem will inspire you to join the quest for community-led solutions towards brighter future for the deeply intertwined human-water systems."

Dr Ana Mijic, *Reader in Water Systems Integration and Director of the Centre for Systems Engineering and Innovation, Imperial College London, UK*

"In a world where we are often siloed by our personal preferences and careers, this book brings together perspectives from the disciplines of positive psychology, engineering, and science. It gives a good insight into how helping to protect and enhance biodiversity, and the environment can have a really positive impact on your personal wellbeing."

Barry Deane, *CEO, National Federations of Group Water Schemes, Ireland*

"This thought-provoking and timely book delves into the profound connections between positive psychology, wellbeing, and pro-environmental endeavours. By illuminating how our mental health and connection to nature are inextricably bound, the authors offer invaluable insights and practical strategies for fostering the synergies between individual flourishing and collective environmental stewardship. The book is a must-read for anyone seeking to promote personal and planetary thriving in the face of global challenges."

Dr Andrea Giraldez-Hayes, *Chartered Psychologist and Staff Wellbeing Manager at Imperial College London, UK*

"This book is an absolute must-read for anyone seeking inspiration to build a better world for themselves and for our planet. The authors present an optimistic and transformative perspective on pro-environmental efforts through the lens of positive psychology. Enriching the discourse with empirical evidence and firsthand testimonies, they foster a holistic approach to environmental activism that promises enhanced wellbeing for individuals, families, and communities alike."

Dr Pawel Fortuna, *Katolicki Uniwersytet Lubelski, Experimental Psychology Department, Lublin, Poland*

Positive Psychology and Biodiversity Conservation

This book reveals how pro-environmental actions can boost individuals' and communities' psychological, social, and emotional wellbeing, resulting in positive environmental changes.

Pro-environmental actions are often viewed as being motivated by anxiety, shame, or anger. However, emerging research indicates that they can also become a source of positive affect, life meaning, engagement, and other wellbeing outcomes. This book turns the current research and practice of pro-environmental action on its head. Drawing from the field of positive psychology, a rapidly developing science of wellbeing, the book explores new perspectives on how researchers and practitioners can influence engagement in pro-environmental initiatives. It provides ways in which individuals passionate about the environment can reframe their feelings and thoughts and allow their newly gained perspective to improve their wellbeing, and outlines approaches to support and encourage those less motivated to engage in pro-environmental actions. The book draws on research from the biodiversity project called Let It Bee, but also looks at examples of other pro-environmental research, such as water conservation, recycling, and reducing the consumption of meat. This book can be used as a guide for changing how stakeholders motivate people to engage in pro-environmental action.

This book will be essential reading for students and scholars of biodiversity conservation, environmental sustainability, ecosystem services, and environmental psychology.

Jolanta Burke (PhD) is a Chartered Psychologist (British Psychological Society) and a Senior Lecturer (US: Associate Professor) at the Centre for Positive Health, RCSI University of Medicine and Health Sciences, Ireland. Dr Burke specialises in the application of positive psychology. Specifically, she researches mechanisms for enhancing psychological flourishing and new wellbeing interventions, such as pro-environmental, nature-based, and lifestyle-medicine-based tools for enhancing psychological flourishing. Her latest research projects include an exploration of the nature-related mechanisms impacting physiological (Heart Rate Variability, electroencephalogram) and psychological wellbeing (funded by the Science Foundation Ireland) and the psychological wellbeing impact of bees on bee-keepers, their families, and the community (funded by the Irish Research Council and Community Foundation Ireland). She has authored over ten books, published over 60 peer-reviewed publications, and written for such newspapers as the *Guardian*, *Irish Independent*, and *Irish Times*. Her research and publications have been featured in such media as *The Economic Times*, Channel News Asia, CNN, and Fox News. She is a regular contributor to *Psychology Today* and The Conversation, where over two million people have read her articles in the last two years, and she was acknowledged by the *Irish Times* as one of thirty people who make Ireland a better place. For more information, go to www.JolantaBurke.com.

Darren Clarke (PhD) is an Assistant Professor in climate change at Dublin City University. His research specialises in environmental psychology, climate change policy, governance, adaptation, and social vulnerability. He has also worked with the local government sector in Ireland to support them in developing and implementing key performance indicators to measure their climate action progress over the period 2020–2029. He also helped organise Ireland's Children and Young People's Assembly on Biodiversity Loss – the first of its kind globally. He currently leads a national climate adaptation research project funded by the Environmental Protection Agency examining the political economy of climate adaptation in Ireland. He is a regular contributor to national and local print media in Ireland, including the *Irish Times*, the *Irish Independent*, and RTE Brainstorm. He also regularly appears as an expert on national television on topics related to sustainability and climate change.

Jimmy O'Keeffe (PhD) is an Assistant Professor of Environmental Systems at Dublin City University. He has over 14 years' experience working on natural capital and ecosystem service research in Ireland, UK, and South Asia collaborating with farmers, developers, regulatory authorities, and governments. Jimmy has authored multiple papers on the use of systems modelling and stakeholder engagement, helping to bridge the gap between engineering and social science. He also helped organise Ireland's Children and Young People's Assembly on Biodiversity Loss – the first of its kind globally. Jimmy's current research focuses on quantifying the societal benefits, particularly around health and wellbeing provided by urban natural capital. He uses interdisciplinary approaches, including stakeholder engagement and systems modelling, to improve our understanding of the links and feedbacks between humans and the natural environment.

Sean Corrigan is a Water and Environmental Engineer. He has over 20 years' experience in the water sector working on projects from catchments to the tap. Through nature-based, and community-focused solutions, the projects that Sean has managed have reduced the need for capital expenditure. They have also been recognised national and internationally winning numerous awards. Sean is also the sustainability director of a children's charity that focuses on the child and parent's physical and emotional wellbeing as well as focusing on biodiversity, upcycling, and carbon sequestration. The charity will be carbon positive by 2027.

Routledge Studies in Conservation and the Environment

This series includes a wide range of inter-disciplinary approaches to conservation and the environment, integrating perspectives from both social and natural sciences. Topics include, but are not limited to, development, environmental policy and politics, ecosystem change, natural resources (including land, water, oceans and forests), security, wildlife, protected areas, tourism, human-wildlife conflict, agriculture, economics, law and climate change.

Creating Resilient Landscapes in an Era of Climate Change
Global Case Studies and Real-World Solutions
Edited by Amin Rastandeh and Meghann Jarchow

Species, Science and Society
The Role of Systematic Biology
Quentin Wheeler

Conservation Concepts
Rethinking Human-Nature Relationships
Kurt Jax

Japan's Withdrawal from International Whaling Regulation
Edited by Nikolas Sellheim and Joji Morishita

Conservation Leadership
A Practical Guide
Simon Black

Positive Psychology and Biodiversity Conservation
Health, Wellbeing, and Pro-Environmental Action
Jolanta Burke, Sean Corrigan, Jimmy O'Keeffe, and Darren Clark

For more information about this series, please visit: www.routledge.com/Routledge-Studies-in-Conservation-and-the-Environment/book-series/RSICE

Positive Psychology and Biodiversity Conservation

Health, Wellbeing, and Pro-Environmental Action

Jolanta Burke, Darren Clarke,
Jimmy O'Keeffe, and Sean Corrigan

LONDON AND NEW YORK

from Routledge

First published 2025
by Routledge
4 Park Square, Milton Park, Abingdon, Oxon OX14 4RN

and by Routledge
605 Third Avenue, New York, NY 10158

Routledge is an imprint of the Taylor & Francis Group, an informa business

British Library Cataloguing-in-Publication Data
A catalogue record for this book is available from the British Library

ISBN: 9781032590394 (hbk)
ISBN: 9781032590400 (pbk)
ISBN: 9781003452676 (ebk)

DOI: 10.4324/9781003452676

Typeset in Sabon
by codeMantra

Contents

List of tables xiii
Acknowledgements xv
Book outline xvii

PART 1
The foundations 1

1 The need for change: an introduction 3
Environmental impacts to health 6
Environmental solutions 6
The role of pro-environment behaviour 7
The need for a whole systems approach 9
Coexistence Management 11
Nature-based thinking (NbT) and nature-based solutions (NbS) 12
Positive psychology 16
Current book's approach 18

2 Pro-environmental actions and wellbeing 22
Defining pro-environmental behaviour 23
Why being environmentally friendly might make us feel good 25
Evolutionary traits: intrinsic versus extrinsic motivations 26
Social norms and values 29
What influences wellbeing in PEB 31
Types of PEB that determine wellbeing 32
Behaviour determinants 32
Individual or collective cultures 33
Socio-demographics of PEBs and wellbeing 35
Encouraging pro-environmental actions for wellbeing 36
Conclusion 38

3 Natural vs urban environments vis-à-vis health and wellbeing 45
Introduction 46
Pollution and mental health 46
Natural versus urban environments 50
Green spaces and mental health 51
The role of nature in shaping wellbeing 52
The need for evidence 56
Conclusion 57

PART 2
Individuals' wellbeing, pro-environmental behaviour and biodiversity 63

4 Environment in the context of illbeing and wellbeing 65
Mental health and environment connection 66
Flourishing in Europe 74
The Bee Well project 75
Conclusion 78

5 Individual emotional wellbeing 81
Emotional wellbeing 82
Broaden-and-Build theory 83
Positive emotions and the environment 85
Resilience 87
Eco-anxiety and positive emotions 89
Life satisfaction 98
Conclusion 99

6 Individuals' psychological wellbeing 104
Engagement 106
Sense of accomplishment 107
Anticipation 109
Hope and optimism 111
Purpose and meaning-making 119
Conclusion 124

7 Mindsets as guideposts for wellbeing 128
Passion 130
Growth mindset 135
Stress mindset 137

Time perspectives and pro-enviro behaviour 140
Maximising vs satisficing 144
Conclusion 145

PART 3
The impact of wellbeing within broader social pro-environmental networks 149

8 **Family's health and wellbeing** 150
Background to family relationship 152
Character strengths 156
Capitalisation 162
Savouring 164
Conclusion 167

9 **Community health and wellbeing** 170
Community and environmental practice 171
Ecosystem model and pro-environmental actions 171
Positivity resonance – emotional contagion 183
Sense of belonging 185
Forgiveness 190
Conclusion 192

10 **The future of pro-environmental actions, health and wellbeing** 195
Character strengths 195
Individuals and communities pulling together in shared goal 197
Health and environmental agendas combined 198
Pro-environmental interventions combined with positive psychology interventions 199
Balancing self-care and passion 200
Final word for now 201

Index 203

Tables

3.1 Pollutants and their impacts on mental health 47
3.2 Urban design principles for improving mental wellbeing 54
8.1 Seven virtues and 24 character strengths 156
8.2 Applying character strengths to pro-environmental behaviour within a family (adapted from Warren & Coughlan, 2016) 158

Acknowledgements

We wish to thank the Irish Research Council and Community Foundation Ireland for funding the Bee Well project, which has inspired this book.

Book outline

Pro-environmental endeavours are often intertwined with emotions such as anxiety, shame, or anger, which frequently capture the attention of researchers and practitioners, delving into phenomena such as eco-anxiety. However, within this emotional terrain, nascent research suggests a transformative potential within these very emotions. It suggests pathways adorned with positive emotions, infused with a sense of purpose, and echoing with engagement, offering the promise of holistic wellbeing.

This publication dares to confront entrenched conventions in pro-environmental research and practice, highlighting a hopeful perspective amidst the evidence. It posits that embracing efforts to cultivate biodiversity not only catalyses positive environmental transformations but also fosters psychological, social, and emotional wellbeing for individuals, their families, and communities.

This approach, grounded in positive psychology (the science of wellbeing), breaks away from conventional norms. It encourages avid environmental advocates to recalibrate their thoughts and emotions, uncovering an increase in personal fulfilment in the process. Additionally, it reaches out to those with less inclination towards environmental activism, spanning the chasm between environmentalism and individual wellbeing.

This book ventures into uncharted territories for researchers and practitioners, with the aim of shaping participation in pro-environmental endeavours. Through positive communication, it celebrates the merits of health and wellbeing as drivers of transformation. Amid empirical validation, it enriches its narrative with findings from interviews with participants in the Let It Bee biodiversity project, vividly illustrating the intertwining of biodiversity enhancement with personal, family, and community wellbeing.

The Bee Well project

In the idyllic landscapes of Ireland, the threat to its waterways from agricultural practices is starkly evident, with incidents like slurry spillage staining the rural beauty and water quality. This harsh reality has led to a drastic decline in biodiversity, with a mere 0.6% of Irish rivers maintaining their pristine status

(DHPLG, 2018). The Irish National Biodiversity Plan echoes this concern, identifying Agriculture, Forestry, Fisheries, and habitat loss as the primary threats to EU-protected habitats' sanctity (NPWS, 2017). More than just compliance with environmental regulations is necessary to stop this deterioration.

Amidst deterioration in water quality, scientific knowledge has advanced in tackling these existential challenges. Yet, translating these scholarly insights into actionable steps at the grassroots level remains a formidable task. Within this complex landscape of aspirations and constraints, the National Federation of Group Water Schemes unveiled its ambitious vision through the National Groundwater Source Protection pilot programme in County Roscommon, Ireland in 2019.

This all-encompassing initiative has brought a multitude of positive changes. Inspired by it, 30 farmers have taken up beekeeping, guided by training and mentorship, fostering a variety of environmentally friendly farming practices and a deep respect for nature. Other aspects of the programme have introduced measures to protect the purity of water bodies, conserve the vitality of bees, reduce the use of pesticides, and promote nature-based solutions to strengthen the protection of waterways. It's an example of conservation, resulting in creating riparian habitats, nurturing the promise of enhanced biodiversity, and the absorption of nutrients from the fertile soils.

At the heart of this initiative is a testament to the unyielding spirit of stewardship, composed of collective effort and unwavering dedication. As the sun bathes the rippling waters in golden hues and the buzz of bees fills the air, there is a glimmer of hope – a hope that amidst the challenges facing our water bodies, the harmonious interaction of nature and humanity can be restored, one beehive, one riparian habitat, one soul at a time.

Below are several examples that transpired because of the Let It Bee initiative:

Bee Farmers: A select group, equipped with beehives and knowledge, tended to their land reverently, fostering a deeper connection with nature.

Children: Environmental stewardship took root in the hearts of the young as tree planting and pesticide-free gardening initiatives flourished, nurturing a symbiotic relationship between youth and nature.

Local Voluntary Groups: A group of mental health patients built "bee hotels" and locals placed them in their gardens, sparking conversations on nature's vital role in the community.

Bee Farmers' Neighbours: Educational efforts enlightened neighbours about the impacts of pesticides on water, advocating for the preservation of bee habitats.

Habitat Creation: Farmers and neighbours collaborated to establish bee habitats, supporting native pollinators, purifying waters, and sequestering carbon.

Amid these narratives, a poignant story unfolded of one farmer gifted with a single beehive, who transformed his land into a sanctuary for biodiversity, inspiring neighbours to follow suit, which resulted in meadows and cleaner water in the area. The Let It Bee project transcends mere water quality improvement; it is a testament to the collective synergy of community collaboration and biodiversity enhancement.

After the Let It Bee project, the Bee Well initiative secured funding from the Irish Research Council and Community Foundation Ireland to investigate its effects on individual farmers' wellbeing, families, and the community. Findings revealed enhancements across various domains of psychological, emotional, and social wellbeing (Burke & Corrigan, 2024; Burke et al., under review), highlighting the significance of such pro-environmental endeavours, not only on the environment but also on human health and wellbeing.

Using honey bees as a catalyst to unite and empower communities is a concept that has been explored previously, as evidenced by various case studies in cities across America. One notable example is Governors Island, a former Army command centre in New York City, which has been transformed into an urban farm with a strong sustainability ethos. This encompassed a range of biodiversity initiatives, including composting, native flower beds, and bee sanctuaries. Governors Island continues to serve as a volunteer-driven hub for urban nature, offering respite from the hustle and bustle of city life. It stands as a compelling example of the manifold benefits of biodiversity enhancement in an urban area.

Similarly, the Let It Bee project stands tall as an embodiment of the harmonious relationship between environmental actions and human flourishing. Readers are invited to explore the outcomes of the impactful Let It Bee project and other tales of environmental dedication. This book seeks to highlight the profound connection between pro-environmental behaviours and biodiversity preservation, portraying them as agents of environmental change and guardians of individual wellbeing.

The journey within these pages navigates the shifting landscapes of perception and motivation in pro-environmental actions. It provides a critical lens through which existing strategies are examined, paving the way for an alternative approach rooted in enhancing wellbeing. As chapters unfold, the concept of wellbeing is redefined, exploring this field from a fresh perspective.

In the culmination of this journey, readers are granted insights into the budding tendrils of research supporting this innovative approach. The aim, profound in its simplicity, is to deepen understanding of the symbiotic relationship between pro-environmental actions and individual wellbeing, a relationship that resounds across the broader ecological canvas, providing hope and resilience in our shared existence.

Book outline

This book is divided into three parts. Part 1 delves into the foundational aspects of the correlation between pro-environmental actions, biodiversity, health, and wellbeing. In Chapter 1, we begin by clarifying the necessity for a shift in our approach to pro-environmental actions. Subsequently, in Chapter 2, we focus our attention on urban areas and associated wellbeing. Chapter 3 delves into the intricate relationship between wellbeing and pro-environmental actions, exploring various moderating factors such as culture and socio-economic status. By the end of Part 1 of this book, you will understand the potential of pro-environmental actions in enhancing wellbeing.

Part 2 of the book focuses on the individual, explicitly exploring concepts derived from positive psychology that aid individuals in enhancing their wellbeing within the realm of pro-environmental actions, including protecting biodiversity. We delve into emerging research in this domain and explore how interventions targeting wellbeing can be adapted to enhance wellbeing and bolster engagement with environmental actions. This section is divided into three chapters: Chapter 4 is an introduction to the individual's wellbeing. We investigate the various positive psychology perspectives on wellbeing and their generic impact on pro-environmental behaviour. Chapter 5 explores emotional wellbeing. Chapter 6 delves into psychological wellbeing. It also discusses mindsets that guidepost wellbeing, meaning choices individuals need to make when pursuing pro-environmental stewardship that can impact their wellbeing positively or negatively. By the end of Part 2, you will understand the positive psychological perspective on wellbeing and how the various elements of wellbeing impact pro-environmental actions and vice versa.

In Part 3 of the book, we focus on the impact of wellbeing within broader social networks. Chapter 7 examines family wellbeing in the context of pro-environmental actions. Chapter 8 delves into the community wellbeing. Finally, in Chapter 9, we briefly offer suggestions regarding the future application of positive psychology in environmental sciences.

We hope you will enjoy this book and find it useful.

Part 1

The foundations

Part 1 of this book serves as a comprehensive introduction to the foundational principles of pro-environmental behaviour and its intricate relationship with health and wellbeing. Our exploration commences with an honest appraisal of the current state of affairs, confronting environmental devastation, a term that refers to the destruction or ruin of the natural environment that plagues our planet and the urgent need for transformative change. Within these pages, we navigate through the myriad pathways charted by researchers and practitioners alike, each offering potential avenues for effecting meaningful change. One such pathway, central to the overarching theme of this book, examines the intersection of positive psychology and emerging positive ecology, illuminating how these disciplines can synergistically contribute to catalysing positive environmental action.

In Chapter 2, we delve into the phenomenon of urbanisation and its far-reaching implications for human health and wellbeing. As our world undergoes rapid urbanisation, we confront the profound consequences it imposes on individuals and communities. Through our exploration, it becomes evident that the speed and scale of urbanisation necessitate proactive environmental measures, urging a collective effort to engage in pro-environmental action. The imperative to mobilise widespread participation in this endeavour highlights the urgency of our times, compelling us to act now.

Moving forward to Chapter 3, we turn our focus to the pro-environmental actions themselves, examining their tangible impacts on wellbeing and health. Through empirical evidence and case studies, we unravel the interconnectedness between environmental stewardship and human flourishing. This means that when we take care of the environment, it reciprocates by providing us with clean air, water, and food, which are essential for our wellbeing. We confront the sobering reality that the environmental crisis we face demands immediate attention and concerted action. Yet, amidst this urgency, we find hope in the transformative potential of pro-environmental behaviour and the preservation of biodiversity.

DOI: 10.4324/9781003452676-1

As we conclude this section of the book, we emphasise the critical juncture at which we stand. The precarious state of our environment demands vigilance and proactive measures to safeguard its future. By embracing pro-environmental action and championing the conservation of biodiversity, we not only chart a course towards a sustainable planet but also nurture the wellbeing and health of present and future generations. The stakes are high, but so too are the opportunities for positive change. It is our collective responsibility, including yours, to heed the call and embark on this journey towards a healthier, more sustainable world.

1 The need for change

An introduction

We are experiencing a significant decline in biodiversity. Globally, the relentless advance of deforestation and land conversion for agricultural and commercial purposes devastates the intricate ecosystems that support life. Yet, the toll exacted upon our natural world extends far beyond the realm of plants and animals, permeating the very essence of our existence and contaminating the waters vital for life.

One of the primary ways humanity has altered the environment is through changes in land use. Over decades, forests have succumbed to swing of the axe, wetlands have yielded to the economic progress, and rivers have given way to the imposing force of dams – all to meet the demands of societal progress and economic growth. However, in this transformation of landscapes, the consequences extend beyond mere reshaping of terrain.

These land-use alterations, orchestrated by humans, represent a paradigm shift where habitats are transformed, and nature's resources are exploited to fulfil civilisation's needs. Yet, the toll is not limited to the ecological realm alone. Instead, it echoes through the intricate web of ecosystem life-sustaining services, disrupting the flow of vitality upon which all living beings rely. Such changes have left an enduring mark on the terrestrial realm, with approximately 75% of land-based environments and nearly 70% of marine ecosystems bearing unmistakable signs of human intervention (Bongaarts, 2019).

As a result, more than a third of the Earth's landmass and nearly three-quarters of its freshwater reservoirs are now dedicated to crop cultivation or livestock farming. In this landscape dominated by human activity, we are confronted with a biodiversity crisis of unparalleled magnitude. Life on our planet is unravelling at a distressing pace, and declining at levels that surpass any in human history (Dasgupta, 2021). This erosion impacts nature's resilience and adaptability, casting shadows of uncertainty and peril over our economies and collective wellbeing (Dasgupta, 2021). Over a century, the abundance of native species on land has decreased by 20%, with the looming threat of extinction hovering over nearly one million species, many of which are on the brink within the next decade (Bongaarts, 2019).

DOI: 10.4324/9781003452676-2

Recent times have witnessed a distressing downward spiral in the quality of rivers, lakes, and groundwater. These surface water bodies serve as crucial drinking water sources and hold an immeasurable amenity value vital for the economy. Sectors like agriculture and tourism, which rely on pristine water sources, face implications from this decline. The Environmental Protection Agency's report on water quality in Ireland from 2013 to 2018 highlights this worrisome trend, revealing a stark decline from 13.4% to 0.6% of sites qualifying as Ireland's highest quality river waters, leaving a mere 20 pristine river sites (EPA, 2019). Similar trends are evident worldwide.

The World Health Organization, in its 2016 report, unequivocally links catchment activities to the protection of drinking water supplies, underlining the intimate connection between safeguarding water resources and ensuring human health and environmental wellbeing. Studies, such as the one by Larkin (2020), have identified the agricultural sector's contribution to point source pollution from livestock and diffuse pollution from pesticide runoff, exacerbating the challenges faced by water quality and biodiversity conservation.

Human activity emerges as the instigator of environmental disruption. Our appetite for progress, marked by industrialisation and urbanisation, has left a mark on the natural world. From the ocean's depths to the vast expanse of the sky, our presence manifests itself as microplastics in oceanic currents and aerosols in the atmosphere.

Driven by necessity and ambition, our species has forged a path paved with consumption and expansion, heedless of the toll on nature's delicate balance. The relentless pursuit of economic prosperity resulted in excessive waste, pollution, and emissions, entrenching patterns of consumption that are not easy to rectify. Yet, our existence, closely linked to the natural world, is also the foundation of our resilience and wellbeing.

Embedded within the complexities of nature lies the concept of Natural Capital, our stocks of natural resources upon which humanity relies for sustenance and survival. From the fertile soils to the crystal-clear streams, these elements form the bedrock of our economy and the cornerstone of our cultural heritage.

Yet, the environmental degradation extends far beyond mere inconvenience, casting a shadow upon the marginalised and the disenfranchised. In the climate change and ecological decline, it is often the most vulnerable among us who bear the brunt of impacts caused by declining environmental quality. The environmental injustice is entwined with the fabric of social and economic inequality.

In the adversity, emerges a developing recognition of the symbiotic relationship between ecological stewardship and human flourishing. Inspired by environmental consciousness, individuals and communities find themselves empowered to enact change, to take care of the world that sustains them. Through the cultivation of pro-environmental behaviours – be it the mindful tending of native flora or the conscientious reduction of pesticide usage – humanity embarks upon a journey of redemption, forging a path towards a future where the wellbeing of both people and the planet are inextricably intertwined.

To counteract this distressing trajectory, the EU Water Framework Directive sets forth its ambition to reverse the tide of degradation. However, the initial cycle of its implementation faltered due to inadequate public and community engagement. The reasons cited for this shortfall were the application of a "one size fits all" and an overly generalised approach (Daly et al., 2016). Article 14 of the Water Framework Directive highlights the importance of stakeholder engagement, emphasising the crucial role of communication in leading environmental change.

In our times, the relentless environmental degradation bears witness to a profound disruption of the natural balance. Human actions have altered at least 70% of global land surface area; there has been an 85% reduction in wetland area since the 1700s while almost 80% of rivers longer than 1,000 km no longer flow freely from source to the sea (Díaz, 2019). This alarming decline, fuelled by habitat loss, fragmentation, and decay, is a testament to humanity's relentless pursuit of progress. Driven by an unyielding desire for advancement, land use transformation, such as deforestation for agriculture and urbanisation, leaves behind a trail of devastation, pushing numerous plant and animal species to the brink of extinction. Concurrently, it disrupts the balance of ecosystems, eroding their capacity to provide essential services – clean air, fresh water, and fertile soil – upon which all life depends (Hansen et al., 2020). This degradation is also exacerbated by pollution, climate change, and overexploitation of natural resources, all of which are largely caused by human activities.

The intensification of agriculture is driven by society's insatiable hunger for plentiful sustenance. However, the widespread use of chemical fertilisers and pesticides results in degradation, contamination of water sources and erosion of biodiversity. Across vast expanses, monoculture farming practices disrupt the genetic diversity within crop species, leaving agricultural systems susceptible to pests, diseases, and the relentless environmental upheaval, exacerbated by the threat of climate change.

Another manifestation of environmental transformation lies in the unrestrained exploitation of renewable and non-renewable natural resources. Humanity's consumption surpasses nature's ability to regenerate, signalling the threat of ecological bankruptcy. Indeed, to maintain our current standard of living, the urgent demand on the planet's resources would require the provision of 1.6 Earths – a stark testament to the unsustainable nature of our lifestyles (Dasgupta, 2021).

Across the globe, mining minerals, harvesting fossil fuels, and extracting raw materials exact a toll upon the very bedrock of existence. Habitat destruction, soil erosion, and water pollution stand as testaments to the greed of human endeavour, threatening the delicate equilibrium upon which ecosystems and the communities that rely upon them depend. Moreover, the extraction and refinement of these resources increase greenhouse gas emissions, impacting climate change that further imperils the integrity of our environment.

Environmental impacts to health

Amidst the backdrop of environmental decline, the impact on human health and wellbeing is profound and far-reaching. Across the crucial trio of air, water, and soil, the infiltration of aerosols, toxic compounds, and heavy metals casts a dark shadow over our existence. In the midst of polluted air, every breath impacts our wellbeing. This toxic exchange not only breeds respiratory ailments and cancers but also results in debilitating neurological disorders. Once-pure air now carries the burdens of industrial excess, disrupting the balance of our physical and physiological wellbeing.

On a global scale, a staggering 74% of annual deaths are attributed to non-communicable diseases, including cancers, cardiovascular disorders, diabetes, and chronic lung ailments (WHO, 2024). While individual lifestyle changes are deemed vital to mitigate this alarming trend, the environmental degradation is a significant contributing factor. Pollution, in particular, is closely linked to the prevalence of non-communicable diseases (Fuller et al., 2018), with countries like Canada, France, Germany, the UK, and the USA leading in pollution levels and disease prevalence. Pollution also serves as a significant precursor to lung diseases, including the threat of lung cancer (Schraufnagel et al., 2019). Thus, the repercussions of environmental deterioration on human health represent an unprecedented crisis, which will impact the wellbeing of future generations.

Similarly, once-pristine waters are now contaminated, unchecked pathogens give rise to waterborne diseases that often afflict the most vulnerable populations. In the developing countries, where inadequate sanitation looms large, millions fall victim to illnesses like cholera and dysentery that are easily preventable.

Across continents, the toll of environmental degradation highlights the interconnectedness between our planet's fate and the wellbeing of its inhabitants. Faced with this silent onslaught, we urgently call for immediate and concerted action to safeguard our environment's sanctity and preserve the health and wellbeing of future generations.

Environmental solutions

With a rising tide of climate change consciousness, the environmental movement is growing. As it permeates the collective consciousness, it begins to reshape the landscape of our policies and regulations. Driven by public outcry and economic pragmatism, this evolution signifies a growing realisation that safeguarding our environment is both a moral and practical.

However, despite the urgency, an array of obstacles obstruct our path. These barriers include deeply ingrained societal norms prioritising economic growth over environmental sustainability, complex regulatory frameworks often favouring industry interests, and a lack of public awareness and engagement on environmental issues.

At the same time, the emerging awareness of our profound interconnectedness with the natural world sparks a collective call to action to confront the environmental threats surrounding us. Governments, businesses, and civil society are awakening to the urgent need for environmental stewardship, each playing a crucial role in reducing their ecological footprint and promoting sustainability. Nevertheless, challenges persist and need to be addressed promptly.

To appear environmentally conscious, organisations often resort to deceptive practices to portray themselves as eco-friendly – a phenomenon termed "greenwashing". This is like a wolf in sheep's clothing, where corporations seek to assuage their conscience while exploiting nature under the guise of environmental benevolence.

Hence, a fundamental shift is required in our lifestyles and how we manage resources. This transformation must originate from large conglomerate companies, government bodies and extend to grassroots initiatives – to ensure the widespread adoption of environmentally friendly behaviours. Each individual has a role to play in this collective endeavour, whether reducing their carbon footprint, supporting sustainable businesses, or advocating for more robust environmental policies. Only through this collective endeavour can we envision a sustainable future where the harmony between humanity and nature prevails.

People often operate within silos, with environmentalists focusing solely on biodiversity while others concentrate on water or climate issues. Addressing these environmental challenges requires behavioural change, yet psychologists are seldom involved in project planning. Similarly, psychologists rarely engage with environmentalists or engineers to develop projects that positively impact participants' wellbeing. Imagine the potential if everyone collaborated and worked together synergistically.

The role of pro-environment behaviour

Based on insights from a comprehensive meta-analysis conducted in Bamberg and Möser (2007), the determinants of pro-environmental behaviour hinge prominently on factors associated with knowledge and awareness. The authors propose a nuanced perspective encapsulated in a trilogy of questions, revealing the intricate dynamics influencing environmentally conscious decisions.

The primary focal point identified by Bamberg and Möser is the axis of knowledge and awareness. This foundational understanding is the bedrock upon which individuals base their pro-environmental choices. However, the authors introduce a triad of questions that delve deeper into the decision-making process's psychological intricacies. These questions evaluate the positive and negative personal consequences of selecting a pro-environmental option over alternatives. Additionally, individuals weigh the perceived difficulty associated with the performance of the pro-environmental choice compared to other available options. Crucially, the inquiry extends to determining whether moral obligations exist to advocate for adopting the pro-environmental option.

As articulated by Bamberg and Möser, this triumvirate of questions effectively represents the domains of attitude, perceived behavioural control, and moral norm. This holistic framework contributes significantly, explaining approximately 52% of the pro-environmental behavioural intention construct. However, 48% of the construct remains unexplained, signalling unexplored factors.

Furthermore, the authors shed light on the realization that intention, in and of itself, accounts for a mere 27% of the variance observed in the enactment of pro-environmental behaviour. This revelation underscores the need to broaden the scope of exploration, delving into alternative pathways that may influence individuals' pro-environmental behaviours.

The findings beckon researchers and practitioners to venture beyond the conventional terrain of knowledge, attitude, and moral considerations. The unexplained variance in pro-environmental behavioural intention and the limited correlation with actual behaviour suggest an array of uncharted factors and motivations. To comprehensively catalyse pro-environmental behaviours, unravelling these latent influences, possibly encompassing emotional connections, social dynamics, or contextual factors that might operate in tandem with knowledge and moral considerations, is imperative.

While the foundations of pro-environmental behaviour rest significantly on knowledge and moral discernment, the intricate interplay of attitudes, perceived control, and moral norms elucidates only part of the narrative. The considerable unexplained variance prompts a call to explore novel avenues and untapped dimensions that could further enrich our understanding of effectively stimulating pro-environmental behaviours in individuals.

A myriad of influences spurs pro-environmental behaviour. For some, these actions are intertwined with a passion for ecological outcomes and a commitment to creating a better environmental world, ensuring the planet's survival for generations. Others find motivation in aligning with the United Nations' sustainability goals or regional equivalents, embracing a global perspective on environmental responsibility.

Yet, a distinct group engages in pro-environmental behaviours not just for ecological impact but also for the profound effects on their physical and mental well-being, accompanied by a noticeable reduction in anxiety levels. The positive environmental outcomes serve as gratifying by-products for them, enhancing their overall sense of fulfilment. The social aspect further plays a pivotal role as a motivating force that cultivates a shared sense of community among those dedicated to the noble cause of environmental preservation. The drivers behind pro-environmental actions are diverse and far-reaching, and this list is by no means exhaustive.

> The case of planting trees to protect the southern resident Orcas in British Columbia illustrates the interconnectedness of nature. Only 73 of Orcas remain in coastal waters, and their population is in rapid decline. Deforestation has polluted coastal and inland waters due to runoff from denuded land. Factors like increased forest fires, boat pollutants, erosion from unstable riverbanks, and rising river temperatures have drastically reduced the population of the orcas' primary food source, the chinook salmon. Each Orca consumes approximately 30,000 salmon annually.
>
> Local communities have devised a plan to save the Orcas by planting trees along riverbanks. These trees stabilise the riverbanks, preventing sediment loss, and serve as riparian zones to capture pollutants. They also provide habitat and food for fish while shading and cooling the water, thereby creating optimal spawning grounds for salmon.
>
> Those involved in the project take great pride in their efforts, viewing it as a noble endeavour to protect the environment and preserve biodiversity.

However, another group finds solace in the connection between environmental actions and personal wellbeing – an amalgamation that yields ecological dividends and profound psychological benefits, alleviating anxiety and fostering a sense of fulfilment. The social dimension further amplifies this ethos, forging community bonds among those devoted to the noble cause of environmental conservation. Indeed, the drivers behind pro-environmental actions are manifold, diverse, far-reaching, and contributing to the collective endeavour of safeguarding our planet for generations yet unborn.

The need for a whole systems approach

In conventional environmental research, the emphasis often centres on viewing the interaction between humans and the environment through the lens of conflict. In recent times, the field of ecology has experienced a notable shift, steering away from portraying the connection between humans and the environment as an inherent conflict. Instead, innovative perspectives like Coexistence Management (Chapron & Lopez-Bao, 2016), Convivial Conservation (Bucher & Fletcher, 2019), and Nature-based Thinking (Randrum et al., 2020) have emerged. These concepts advocate for strategies that promote a balanced and harmonious coexistence between humans and nature without jeopardizing the latter's well-being.

A particularly intriguing addition to this evolving landscape is the concept of "positive ecology," introduced to infuse a positive psychology perspective into ecology by Buijs and Jacobs in 2021. Positive psychology is the science of what is right with people rather than wrong. As such, it shifts towards a more optimistic and collaborative ethos in our relationship with the environment.

Human impacts on the environment are complex and multifaceted with far-reaching implications for the health and well-being of both ecosystems and human societies. By understanding the interconnectedness between our actions and the environment, we can begin to address the root causes of environmental degradation and work towards a more sustainable and equitable future.

Our approaches to understanding and solving the environmental crises have been discipline specific; ecologists have focused on impacts to biodiversity, climate scientists on changing weather patterns, policy makers on a particular piece of legislation. Yet all aspects of natural environment are intrinsically connected. Change, in all forms can propagate throughout the system with affects ranging from insignificant to catastrophic. It is not possible to develop realistic or sustainable solutions to environmental problems without acknowledging these interconnections and taking them into account when making decisions. Only through a "systems thinking" approach – a holistic lens that embraces the intricate web of connections and feedback loops between humans and nature – can we hope to untangle the intricacies of environmental crises and forge pathways towards meaningful solutions.

Systems thinking provides a structured approach to understand complex problems, providing a shared whole system view which allows us to understand dependencies and consider different perspectives (Mijic, 2021). This provides us with an approach to understand the many multiple and complex interactions that exist between humans and the natural environment, allowing us to collaboratively develop more coherent management options and policies. It emphasises the recognition of patterns, feedback loops, and emergent properties within systems, considering both the internal dynamics and the external environment. Drawing from disciplines, such as ecology, biology, and engineering, systems thinking helps uncover the underlying structures that influence system behaviour and enables more effective problem-solving and decision-making. As Donella Meadows, a prominent systems thinker, noted in her book "Thinking in Systems: A Primer", understanding systems is crucial for addressing interconnected global challenges and promoting sustainable solutions (Meadows, 2008).

Societally, collective reluctance to make a change comes down to a number of factors, often rooted in economic, practical, and social constraints. For many, the threats posed by large scale challenges such as climate change are existential – too large to consider in our daily lives where the primary concern for most is putting food on the table or paying that next bill. Making pro-environmental changes are often considered a luxury – something for the privileged, or the devoted and not something we can take part in without significantly impacting our quality of life. There are also social constraints where pro-environmental behaviour is looked down upon, ridiculed or opposed. This is often exacerbated by polarising media and political discourse. Therefore, our most significant societal challenge is our willingness to engage in pro-environmental actions.

Coexistence Management

Coexistence Management (Chapron & Lopez-Bao, 2016) represents a game-changer in how we view and address the complex dance between humans and the environment. While conventional narratives have often portrayed this interaction as a conflict, emphasising the detrimental effects of human activities on delicate ecosystems, Coexistence Management offers a radical departure carving a path towards harmony.

This revolutionary approach redefines our relationship with the environment, acknowledging the pivotal role of human agency. At its essence, Coexistence Management embodies a philosophy of equilibrium – a balance where human societies and ecological systems converge in mutually beneficial outcomes. It encourages an exploration of innovative strategies that mitigate negative impacts while nurturing positive interactions. Thus, striking a harmonious balance between human necessities and environmental imperatives. However, it is essential to acknowledge that implementing coexistence management can be complex and may face resistance from those accustomed to more traditional approaches. It requires a shift in mindset and a willingness to embrace change.

Its remarkable versatility sets Coexistence Management apart. It is manifested in urban planning, wildlife conservation, or agricultural practices, this paradigm surpasses disciplinary barriers, building bridges between conservation goals and human endeavours by embracing the interplay between social and ecological systems. For instance, Coexistence Management has been used in urban planning to design green spaces that promote biodiversity while meeting the community's needs. In wildlife conservation, it has protected endangered species while respecting the rights of local communities. Moreover, agriculture has led to the development of sustainable farming practices that minimise environmental impact. The solutions associated with Coexistence Management, therefore, increase both environmental sustainability and human wellbeing.

Coexistence Management embodies a forward-thinking approach, challenging the age-old narrative of conflict between humans and the environment. It paves the way for a more symbiotic relationship, where elements coexist and flourish. As our understanding of ecological dynamics evolves,

Coexistence Management stands points us toward a future where environmental conservation and human societies thrive in harmony.

Nature-based thinking (NhT) and nature-based solutions (NbS)

In the interplay between human civilisation and the natural environment, systems thinking is a pivotal tool through which we can understand the complexities of human-nature interaction. However, alongside this intellectual framework, NbT and NbS stand ready as transformative allies, offering a pathway towards resolution. The 2030 Agenda for Sustainable Development, a visionary blueprint established by the United Nations in 2015 and embraced globally, not only highlights the imperative to address poverty, inequality, and the preservation of our planet but also places a significant emphasis on the role of NbT and solutions in our collective global efforts (UN, 2015).

Traditionally, engineering solutions have provided a straightforward approach to socio-environmental challenges – a mindset rooted in human dominance over nature. Yet, as we confront the limitations of this paradigm in the face of climate change and biodiversity loss, a shift is underway. We now realise that it is practical and financially prudent to shift our perspective and work in harmony with nature, preserving the delicate balance of ecosystem services. This change in perspective marks a significant evolution in our approach to socio-environmental challenges.

Nature-based solutions (NbS) emerge at the forefront of this evolutionary shift – an ethos encapsulated in a vision to harness the wisdom of nature to address multifaceted challenges. These solutions, inspired and nurtured by the natural world, offer many environmental, social, and economic benefits while enhancing resilience and fostering synergy among sustainable development goals (Seddon et al., 2021).

NbT, at its core, signifies a departure from anthropocentric worldviews, embracing a holistic ethos that positions humanity within the fabric of life rather than above it. Central to this paradigm is the concept of bio-inspiration or biomimicry – a reverent process of drawing inspiration from nature's design principles to tackle human challenges (Bianciardi, 2023). Furthermore, NbT highlights the significance of ecological literacy – a profound understanding of ecosystems and our role within them, which is crucial for fostering informed decision-making and ecological stewardship.

A cultural shift is not just desirable but imperative in advancing NbT. Progress must be reimagined beyond mere economic growth towards a holistic understanding of wellbeing that encompasses ecological harmony and spiritual fulfilment. Education emerges as a linchpin in this endeavour – an instrument to instil ecological literacy and nurture environmental stewardship from an early age.

Governments, businesses, and civil society organisations wield considerable influence in fostering an environment conducive to NbT. Their policies prioritising sustainability, investments in green technologies, and collaborations among diverse stakeholders are crucial in catalysing this transformative shift. Furthermore, integrating biophilic design principles in urban planning and architecture offers a tangible manifestation of this ethos – a fusion of human habitat with the rhythms of nature, nurturing a sense of belonging and stewardship.

In our collective journey towards sustainability, NbT stands as a guiding light – a testament to the boundless ingenuity and resilience inherent in the natural world. As we chart a course towards a more harmonious coexistence, let us heed the wisdom of nature, embracing its ethos of interconnectedness and reverence, as we forge a path towards a future where humanity and the natural world thrive together.

Community-based actions

"Look at them; they are all smiling; they look so happy", remarked Sean Clerkin of the *National Federation of Group Water Schemes* while viewing the photographs capturing farming families' camaraderie for the Let It Bee project. Each family dedicated their time and portions of their cherished lands to enhance biodiversity, safeguard local water sources, and spark hope in climate action amidst the enchanting landscapes of West Ireland. Yet, beyond the tangible benefits relating to environmental protection, the sheer joy emanating from their efforts was impossible to overlook.

As we pondered over these snapshots, a revelation emerged that challenged a prevalent misconception. The idea that environmental stewardship burdens individuals, demanding sacrifice and leading to adverse behavioural changes, clashed with the scenes before us. Contrary to expectations, the faces of these farmers reflected genuine contentment about their participation in the Let It Bee project. Perhaps this is what the future should look like for many of us interested in environmental protection.

Over 70% of the Ireland's EU-designated habitats bear the brunt of agricultural impact. The loss of hedgerows and marginal lands to monoculture is a stark testament to the decline of cherished species like butterflies and bees. However, amidst these challenges, a shift in attitudes towards biodiversity is palpable. Once viewed as a moral and ethical obligation, nature's protection now resonates across various sectors of the economy. The tourism industry mainly showcases nature's allure, with overseas visitors drawn to Ireland's pristine beauty, each euro spent echoing the harmony between biodiversity, ecosystem function, and human wellbeing.

The International Union for the Conservation of Nature (IUCN) champions NbS as potent tools against global environmental damage. Water security, biodiversity preservation, and ecosystem resilience are pivotal battlegrounds where nature's inherent wisdom offers a formidable defence. These principles lay the groundwork for the Let It Bee transformative action.

Commissioner Virginijus Sinkevičius of the European Commission praises NbS as the vanguard in the fight against climate change. These solutions, he argues, are not just remedies but catalysts for empowerment, community resilience, and economic prosperity. Investing in nature becomes prudent and imperative for safeguarding our collective future.

One truth emerges resoundingly clear in the motivations driving individuals towards biodiversity preservation: the interplay between biodiversity and environmental protection is inseparable. Whether fuelled by moral imperatives, economic incentives, or ecological stewardship, safeguarding biodiversity is a universal call to action, guiding us towards a sustainable and harmonious future.

In the delicate dance of nature, honeybees and their fellow pollinators traverse human-altered landscapes for sustenance. However, this can be met with peril as chemical pesticides loom large, casting a shadow over their existence. The ramifications of these toxins ripple through ecosystems, leading to the distressing loss of wild pollinators and bees alike.

The echoes of this ecological crisis echo through the corridors of the UK government's DEFRA's pollinator strategy. Here, habitat loss emerges as the principal antagonist, jeopardising pollinators and the sanctity of watercourses. As agricultural runoff seeps into riparian zones, the once-pristine waters face an existential threat of increased contamination.

Amidst this crisis, bees transcend their humble status, emerging as potent symbols of biodiversity's plight. Their decline is spurring a groundswell of action across communities. In Ireland, a humble beehive supports conservation measures as part of the Let It Bee project, united disparate voices and improved environmental outcomes in the process. Here, beekeepers and laypersons converged, their commitment to safeguarding vital pollinators bridging gaps and fostering unity.

The Let It Bee project contributed significantly to the water source protection in the area. Combining scientific rigour with a spirit of collaboration, it brought in understanding and community's collective responsibility. Conversations with farmers revealed the stark reality of environmental degradation, each highlighting the profound challenges those tied to the land face daily. However, farmers also charted a new course rooted in nature's wisdom, empowered by knowledge and guided by a shared vision of stewardship. They reclaimed hedgerows, cultivated apple orchards, and embraced rewilding. All their efforts demonstrated the transformative power of grassroots action.

In rural communities' every sapling planted, every hive tended, resonated with the promise of a brighter tomorrow. As the tendrils of community empowerment spread, they intertwined with the ethos of collective purpose, bringing people together. Each participant stewarded change, each action became a testament to the transformative potential of unity.

Why did joy radiate from their smiles amidst the sacrifice of time and the nurturing of precious earth? The answer came through research revealing the Let It Bee project's impact on personal, familial, and community wellbeing – the Bee Well project (Burke & Corrigan, 2024). Within the crucible of environmental stewardship lay a harmonious symbiosis, where the fortunes of the land and the people who tend to it ascend hand in hand. A triumph, indeed, where the environment and its stewards reap the rewards – a victory shared by all.

Could this revelation, drawn from the Irish fields, serve as a potent tool in the quest for biodiversity enrichment? Might the promise of heightened wellbeing kindle enthusiasm as we seek to safeguard our natural heritage? Contemplating this, the authors, each with their unique perspectives – two from the realm of engineering, the other two in psychology – pondered the manifold implications: different motivations, different benefits, yet a convergence upon shared goals.

Human nature inclines towards that which touches upon everyday existence. Personal connections are part of our lives, whether through financial worries, concern for loved ones' health, or the joys of family and community bonds. We must provide individuals with personal incentives to rally them to the cause of biodiversity, water protection, and climate resilience. This vision stirs the imagination and ignites the spirit. This vision also instigates change.

While this narrative focuses on biodiversity, it is essential to highlight the interconnectedness between biodiversity, water, and climate. Here lies an intricate interdependency – protecting one element yields dividends for another, and harm inflicted upon one echoes throughout the system. In this theme we endeavour to weave the human thread into this intricate area, highlighting the positive ripples that emanate from nurturing biodiversity – an element too often overlooked in the annals of environmental discourse.

Critical role of communication

Reflecting on the inception of his journey into the realm of Civil Engineering, one of the authors (Sean) reminisces about his confident stride into the lecture halls of Sligo Regional Technical College in the late 1980s. Eager to embark on a career designing bridges, erecting water treatment plants, and laying roads, the focus was squarely on the tangible aspects of infrastructure. However, an unexpected emphasis on the importance of communication, courtesy of Joe Cox, the Head of Engineering, challenged the author's preconceptions. In an era where engineers were perceived as builders, not communicators, Cox's insistence on the significance of effective communication left a lasting impression, paving the way for a broader understanding of the interplay between engineering and community engagement.

The Head of Engineering, Joe Cox, emerged as more than a lecturer; he was a visionary, a beacon of wisdom who illuminated the path for his students. In an era where the art of communication was often overlooked in engineering curricula, Cox stood as a testament to the transformative power of words.

His eloquence and knack for articulation transcended the confines of conventional teaching. Cox instilled in his students the profound realisation that effective communication was an ancillary skill and the cornerstone of success. Beyond the technical mechanics of a project lay a realm where connections were forged and lives were touched.

Cox's teachings resonated deeply with his students, enduring long after they had left the halls of academia. In their professional lives, his words echoed with ever-increasing relevance. They learned that true mastery of communication could bridge divides, unite communities, and inspire collective action towards a shared vision of progress. As such, environmental actions rely on effective communication.

In the area of effective communication, listening emerged as a paramount virtue. The ability to convey ideas in a manner that resonated with the listener became the hallmark of a skilled communicator. Such individuals possessed the rare gift of drawing others into their orbit, kindling a sense of ownership and commitment to a cause.

As Cox's lessons demonstrated, the imperative of unifying engineering and psychology became apparent. Effective communication was not merely a tool but a bridge that spanned disciplines, fostering understanding, empathy, and collaboration to pursue a brighter future. Today, decades later, his words continue to lead the field.

Positive psychology

Positive psychology is an exploration of the positive aspects of human experience and flourishing. Departing from the tradition of psychology serving mainly as a support for pathological states and treating mental illness, positive psychology embarks on a voyage into the intricacies of human potential and wellbeing. It casts a discerning gaze upon the facets that elevate existence from mere subsistence to a state of thriving, epitomising optimal human functioning. By doing so, it lays the foundations towards a mentally healthier society.

Venturing on this journey into positive psychology demands dispelling misconceptions that prevent seeing its true value. Positive psychology does not advocate for naive optimism or overlooking life's complexities. Martin Seligman, who pioneered this movement, cautioned against the term "positive", as it oversimplifies the profound intricacies of human existence that include positive and negative experiences, as well as everything in between. Positive psychology, therefore, embodies a nuanced discipline – a pursuit of balance. Given the prevalence of pathologies in psychology, it encourages researchers and practitioners to consider the positive side as well as the negative. This is further exemplified by the second and third waves of positive psychology (Ivtzan & Lomas, 2016; Lomas et al., 2020) that emphasize the need to explore the complexities of human experience, positive side to negative experiences and vice versa. As such, this comprehensive quest encompasses the entire range of human experience, from the depths of adversity to the brilliance of joy.

The roots of positive psychology delve deep into history, intertwining with the psychological scholarship dating back decades. Early explorations into self-esteem, altruism, and autonomy laid the fertile groundwork for the rise of the positive psychology movement in 1998. This seismic shift accelerated the exploration of research relating to human flourishing, emphasising wellbeing, resilience, and the transformative power of individual agency. Through this evolution, the canvas of psychological enquiry expanded, ushering in a renaissance of understanding and embracing the human capacity for growth and fulfilment.

Our innate bias towards negativity (Baumeister et al., 2001), shape perceptions and paradigms across diverse disciplines. Yet, positive psychology challenged the entrenched narratives of deficit and despair. It aimed to strike a delicate balance, acknowledging the evolutionary imperatives driving our preference for negativity while focusing on helping individuals become the best versions of themselves.

Thus, positive psychology emerges not only as a field of study but also as a clarion call – a call to delve deeper, to transcend the limitations of deficit-based paradigms, and to embrace the aspects of human flourishing. By illuminating the pathways to wellbeing and resilience, positive psychology urges humanity to embark on a journey of self-discovery – a voyage into the limitless realms of human potential and possibility.

Positive ecology

Positive ecology weaves together the realms of ecological understanding and human flourishing. Departing from the conventional paradigms that view conservation in isolation, positive ecology embodies a holistic approach that highlights the symbiotic relationship between humanity and the natural world.

At its essence, positive ecology transcends the mere preservation of ecosystems; it delves into the profound influence of nature on human wellbeing. In the ground-breaking research by Buijs and Jacobs (2021), positive ecology emerges as a pathway to nurturing wellbeing – a testament to the reciprocal bond between environmental health and human vitality. This fresh perspective recognises that a thriving environment sustains the planet and elevates our souls, fostering positive emotions, meaningful engagement, and a deep sense of purpose. Buijs and Jacobs delineate three pillars that underpin positive ecology:

1 Positive Emotions: Positive ecology highlights avenues to mental wellbeing by investigating how our interactions with nature evoke positive emotions.

2 Engagement: Individuals cultivate a profound sense of involvement through active connection and participation with the environment, fostering a deeper connection with the natural world.
3 Meaning: The environment acts as a crucible for human purpose and meaning, providing an opportunity for individuals to discover its profound significance.

These pillars, rooted in theories of authentic happiness, serve as guiding stars amidst the vast expanse of ecological enquiry. However, the landscape of wellbeing research continues to evolve, unveiling additional facets crucial to human flourishing.

In human-wildlife interactions, applying these pathways holds promise for enriching our understanding of the intricate dynamics between humanity and the animal kingdom. Buijs and Jacobs highlight three potential impacts of applying these pathways:

1 Organising Existing Research: By systematically structuring current research on human-wildlife interactions, we better understand these encounters and identify gaps in our knowledge.
2 Revealing Unidentified Benefits: Delving into these connections may unveil previously unrecognised benefits, shedding light on positive outcomes that have eluded previous research endeavours.
3 Unravelling Mechanisms: By dissecting the intricate mechanisms underlying the worth and reward of wildlife experiences, we gain insights into the psychological, emotional, and cognitive processes that imbue these interactions with profound meaning.

Moreover, the authors introduce the concept of a potential feedback loop within this dynamic – a symbiotic relationship wherein positive experiences with wildlife contribute to individual wellbeing, fostering a more profound commitment to environmental conservation.

By understanding these pathways, we unlock the door to a deeper understanding of the human-wildlife connection, embrace positive experiences in nature, cultivate wellbeing, and nurture a more environmentally conscious society.

Current book's approach

Within these pages, we champion an environmental conservation approach that cherishes equilibrium. We refrain from turning a blind eye to the challenges and adversities associated with our interactions with the environment. Instead, we implore individuals to confront their feelings of eco-anxiety or similar existential ponderings, acknowledging their presence while also embracing the positive aspects of such encounters. It is the balance of both that makes a positive difference.

Research conducted in the UK reveals that over 30% of individuals navigate through a mixture of negative and positive wellbeing elements simultaneously, indicating that positive and negative wellbeing is independent from each other (Huppert & Whittington, 2003). This suggests that while some may grapple with heightened levels of anxiety, they concurrently experience positive relationships or possess a profound sense of purpose and meaning in life. These facets of wellbeing serve as a helping hand, aiding individuals in coping with their anxieties or, in some instances, facilitating resilience in the face of adversity, thereby helping them alleviate their mental health challenges. This intricate interplay between these contrasting facets, illustrates that they are not solitary entities but coalesce and interact synergistically. As such, they need to be explored individually, which is what this book aims to do.

Within the confines of this volume, we embark on a comprehensive journey delving into the myriad positive dimensions of our wellbeing within the realm of pro-environmental actions. We advocate for individuals to explore these dimensions and actively engage in the suggested wellbeing activities. However, it is crucial to bear in mind that our experiences, whether positive or negative, need to be embraced and accepted. This way, we can practice positive psychology approaches alongside our current experiences (be it eco-anxiety, frustrations, etc.) without denying their existence.

References

Arbor Day Foundation. Wait, Killer Whales Need Trees? | Stories at arborday.org (Accessed: 29 April 2024).

Bamberg, S., & Möser, G. (2007). Twenty years after hines, Hungerford, and Tomera: A new meta-analysis of psychosocial determinants of pro-environmental behavior. *Journal of Environmental Psychology*, 27, 14–25. https://doi.org/10.1016/j.jenvp.2006.12.002

Baumeister, R. F., Bratslavsky, E., Finkenauer, C., & Vohs, K. D. (2001). Bad is stronger than good. *Review of General Psychology*, 5(4), 323–370. https://doi.org/10.1037/1089-2680.5.4.323

Bianciardi, A., Becattini, N., & Cascini, G. (2023). How would nature design and implement nature-based solutions? *Nature-Based Solutions*, 3, 100047. https://doi.org/10.1016/j.nbsj.2022.100047

Bongaarts, J. (2019). IPBES, 2019. Summary for policymakers of the global assessment report on biodiversity and ecosystem services of the Intergovernmental Science-Policy Platform on Biodiversity and Ecosystem Services. *Population and Development Review*, 45(3), 680–681. https://doi.org/10.1111/padr.12283

Buijs, A., & Jacobs, M. (2021). Avoiding negativity bias: Towards a positive psychology of human–wildlife relationships. *AMBIO - A Journal of the Human Environment*, 50(2), 281–288. https://doi.org/10.1007/s13280-020-01394-w

Büscher, B., & Fletcher, R. (2019). Towards convivial conservation. *Conservation and Society*, 17(3), 283–296. https://doi.org/10.4103/cs.cs_19_75

Chapron, G., & López-Bao, J. V. (2016). Coexistence with large carnivores informed by community ecology. *Trends in Ecology & Evolution*, 31, 578–580. https://doi.org/10.1016/j.tree.2016.06.003

Daly, D., Archbold, M., & Deakin, J. (2016). Progress and challenges in managing our catchments effectively. *Biology and Environment: Proceedings of the Royal Irish Academy*, 116B(3), 157–166. https://www.jstor.org/stable/10.3318/bioe.2016.16#metadata_info_tab_contents

Dasgupta, P. (2021). *The economics of biodiversity: The Dasgupta review*. HM Treasury. https://assets.publishing.service.gov.uk/government/uploads/system/uploads/attachment_data/file/962785/The_Economics_of_Biodiversity_The_Dasgupta_Review_Full_Report.pdf

Díaz, S., et al. (2019). Pervasive human-driven decline of life on Earth points to the need for transformative change. *Science*, 366(6471), eaax3100. https://doi.org/10.1126/science.aax3100

Environmental Protection Agency Ireland. (2019). The Environmental Protection Agency's report on water quality in Ireland from 2013–2018. https://www.epa.ie/pubs/reports/water/waterqua/Water%20Quality%20in%20Ireland%202013-2018%20(web).pdf

Fuller, R., et al. (2018). Pollution and non-communicable disease: time to end the neglect. *The Lancet. Planetary Health*, 2(3), e96–e98. https://doi.org/10.1016/S2542-5196(18)30020-2

Hansen, T., et al. (2020). A case study of urban design for wellbeing and mental health in Adelaide, Australia, 6(7). https://www.urbandesignmentalhealth.com/journal-6-adelaide.html

Huppert, F. A., & Whittington, J. E. (2003). Evidence for the independence of positive and negative well-being: Implications for quality of life assessment. *British Journal of Health Psychology*, 8(Pt 1), 107–122. https://doi.org/10.1348/135910703762879246

Larkin, C. (2020). Optimising water quality returns from the reform of the Common Agricultural Policy. Report to An Fóram Uisce|The Water Forum.

Lomas, T., & Ivtzan, I. (2016) Second Wave Positive Psychology: Exploring the Positive–Negative Dialectics of Wellbeing. *Journal of Happiness Studies,* 17, 1753–1768. https://doi.org/10.1007/s10902-015-9668-y

Lomas, T., Waters, L., Williams, P., Oades, L. G., & Kern, M. L. (2020). Third wave positive psychology: broadening towards complexity. *The Journal of Positive Psychology*, 16(5), 660–674. https://doi.org/10.1080/17439760.2020.1805501

Meadows, D. H. (2008). *Thinking in systems; A primer*. Chelsea Green Publishing.

Mijic, D. A. (2021). *Systems water management for catchment scale processes: Development and demonstration of a systems analysis framework*. Environment Agency.

Randrup, T. B., Buijs, A., Konijnendijk, C. C., & Wild, T. (2020). Moving beyond the nature-based solutions discourse: Introducing nature-based thinking. *Urban Ecosystems*, 23, 919–926. https://doi.org/10.1007/s11252-020-00964-w

Schraufnagel, D. E., et al. (2019). Air pollution and noncommunicable diseases: A review by the FORUM of International Respiratory Societies' Environmental Committee, Part 2: Air Pollution and Organ Systems. *Chest*, 155(2), 417–426. https://doi.org/10.1016/j.chest.2018.10.041

Seddon, N., Smith, A., Smith, P., Key, I., Chausson, A., Girardin, C., House, J., Srivastava, S., & Turner, B. (2021). Getting the message right on nature-based solutions to climate change. *Global Change Biology*, 27(8), 1518–1546. https://doi.org/10.1111/gcb.15513

United Nations. (2015). Transforming our world: The 2030 Agenda for Sustainable Development (2030). Resolution Adopted by the General Assembly on 25 September 2015, 42809, 1–13.

WHO. (2016). *Protecting surface water for health, identifying, assessing, and managing drinking water quality risks in surface water catchments* (B. Rickert, I. Chorus, & O. Schmoll, Eds.). World Health Organization.

WHO (2024). *Non-communicable diseases.* World Health Organization. https://www.who.int/news-room/fact-sheets/detail/noncommunicable-diseases

2 Pro-environmental actions and wellbeing

In this chapter, we delve into the intricate interplay between individual wellbeing and pro-environmental behaviour (PEB), focusing on factors that influence its efficacy. By examining these factors, we highlight the need for deeper exploration of this crucial subject matter. Such insights can enrich our understanding and shape our perspectives, facilitating informed discourse and action in environmental stewardship.

> Maggie's journey commenced during her secondary school years, when she heard the impassioned plea of Greta Thunberg to take action that would save the planet. Subsequently, Maggie embarked on a quest to reduce her ecological footprint, and was determined to contribute modestly to preserving our delicate ecosystem.
>
> Maggie's lifelong friend, Carole, accompanied her through their school years. Carole did not quite share Maggie's genuine passion for environmental causes. For Maggie, advocating for environmental issues went beyond mere obligation; it became a source of profound personal fulfilment, infusing her life with a sense of purpose. At the same time, Carole found that her involvement in environmentalism failed to ignite the same sense of excitement within her.
>
> What factors lead to Maggie and Carole's different levels of environmental awareness? What influences contributed to their divergent perspectives? These questions will be explored in more detail in the upcoming chapter.

In the domain of environmental psychology, the dynamic interaction between PEB and human wellbeing is a complex. While prevailing discussions emphasise the symbiotic link between eco-conscious actions and personal wellbeing, dissenting viewpoints introduce nuances, shedding light on the multifaceted factors that influence this relationship.

Advocates champion PEB as a catalyst for wellbeing, praising its potential to cultivate positive shifts in self-perception and instill a profound sense of

DOI: 10.4324/9781003452676-3

purpose. However, sceptics warn against oversimplification, acknowledging the inherent obstacles and sacrifices associated with embracing an environmentally conscious lifestyle.

At the core of this discourse lies the interplay between personal values, intrinsic motivations, and external pressures that intricately defines the relationship between PEB and wellbeing. As we navigate this complex terrain, we grapple with questions of agency, identity, and the intricate cognitive processes that shape individual behaviour.

Our exploration extends beyond theory, as we delve into the real-life experiences of individuals such as Maggie and Carole. Through their stories, we gain insights into the psychological, social, and environmental factors that influence attitudes towards eco-friendly living.

Defining pro-environmental behaviour

In the midst of our fast-paced industrial era; humanity finds itself entangled in a complexity of urgent ecological dilemmas. From the imminent threat of climate disruption to the realities of pollution and resource depletion, our planet bears the scars of our collective apathy (IPCC, 2022). Yet, amidst this turmoil, a ambitious initiatives like the Paris Agreement and the Kunming-Montreal Global Biodiversity Framework aim to mend the wounds we have inflicted upon our Earth.

These monumental efforts cascade from the global arena to regional, national, and local levels, materialising in initiatives, such as the EU Nature Restoration Law, the European Green Deal, and the collaborative endeavours of C40 Cities (C40 Cities, 2019; European Commission, 2021). However, the magnitude of these transformations goes beyond mere policy directives; it demands a profound transformation in societal values and a grassroots resurgence.

PEB emerges as a pivotal point in our efforts to tackle environmental challenges. It is defined as any action aimed at preventing harm to or safeguarding the environment (Steg & Vlek, 2009). PEB encompasses a wide range of activities, from participating in environmental movements to personal conservation practices (Tian & Liu, 2022; Wan & Du, 2022). Recognising the interplay of individual values and daily decisions is crucial to the global mission of repairing the fabric of our natural world.

As we navigate the dynamics between human behaviour and environmental crises, it becomes evident that understanding the motivations behind individual choices is paramount, particularly within the context of personal wellbeing. The success of our collective endeavour to restore and preserve the environment hinges upon our ability to understand and influence these behaviours. By embracing PEB, we not only contribute to the global mission but also pave the way for our own personal growth and more sustainable and harmonious coexistence with our planet.

Global wellbeing

Governments worldwide are becoming more aware of the challenges of relying on traditional measures of wellbeing, acknowledging the inherent shortcomings of relying solely on objective measures of prosperity. For decades, Gross Domestic Product (GDP) has reigned as the dominant yardstick of economic wellbeing, tracing its origins back to the post-war era of the 1940s. However, dissenting voices challenging GDP's adequacy as a measure of overall welfare have gained momentum in recent years, A substantial body of evidence supports these contentions, arguing that GDP fails to capture essential aspects of human flourishing – such as life satisfaction, existential purpose, emotional equilibrium, and environmental vitality (Government of Ireland, 2020; Hickel, 2020). In response, several nations, including Ireland, Canada, Iceland, the United Kingdom, the Netherlands, and New Zealand, have embarked on a journey to adopt more nuanced indicators for assessing national wellbeing, thereby setting a global precedent for a more comprehensive approach to measuring prosperity.

These innovative measures encompass a variety of indices, including the United Nations Human Development Index (HDI), the Genuine Progress Indicator, The Wellbeing of the Nation, and The Living Standards Framework (Government of Iceland, 2011; Government of Ireland, 2020). Moving beyond the narrow confines of economic metrics, these indices traverse the expansive terrain of subjective wellbeing, living standards, economic outcomes, and even GDP. However, despite the growing wealth of comprehensive data, translating these multidimensional wellbeing measures into actionable policies remains a formidable challenge, highlighting the complexity of the task at hand. Objective economic metrics, epitomised by GDP, continue to exert a +disproportionate influence over global perceptions of prosperity, shaping decisions regarding consumption, government expenditure, financial investments, and international trade.

In the midst of this paradigm shift, Bhutan's Gross National Happiness (GNH) index leads the way. Prioritising the dual pillars of wellbeing and happiness in policy formulation, GNH has supplanted GDP as Bhutan's primary measure of progress. Encompassing nine equally weighted domains, including psychological wellbeing, health, education, time use, cultural diversity and resilience, good governance, community vitality, ecological resilience, and living standards, GNH embodies a holistic vision of prosperity, mandating that all government policies align with its principles before implementation. Bhutan's bold experiment in integrating broader measures of wellbeing into policymaking resonates globally.

This exploration into the complexities of contemporary wellbeing is multifaceted, intertwining economic dimensions with the intangible human experience. The sustainability of these wellbeing levels, referring to the ability to maintain or improve current levels of wellbeing over time, hinges on the stewardship of capital stocks, which are the resources and assets that contribute to our wellbeing, essential to our existence – encompassing natural, physical, human, and social domains – with the commitment to pass them on to future generations (Stiglitz et al., 2009).

The inseparable link between human wellbeing and environmental vitality is undeniable. Pursuing the United Nations Sustainable Development Goals (SDGs), a set of 17 global goals designed to be a blueprint for achieving a better and more sustainable future for all, and fostering climate-resilient societies depend on nurturing healthy ecosystems conducive to societal flourishing (IPBES, 2019; IPCC, 2022). Nature-based stewardship emerges as a synergistic complement to conventional sustainability efforts, offering a holistic blueprint for action (Krasny & Delia, 2015). Thus, these multidimensional approaches are essential to future environmental and human wellbeing.

While scholarly discourse may harbour divergent perspectives on the relationship between PEBs and wellbeing that individuals or groups take to reduce their negative impact on the environment – some suggesting a negative or ambivalent connection (Binder et al., 2020) – prevailing research consistently highlights a positive association between PEB and human flourishing (Kaida & Kaida, 2019; Barragan-Jason et al., 2023; Gong et al., 2023). These findings prompt us to delve deeper into the underlying mechanisms driving this connection, unravelling the mystery of why adopting environmentally friendly practices may evoke distinct emotional responses.

Why being environmentally friendly might make us feel good

PEB is not just a trend, but a powerful force that permeates various aspects of our lives. It has the potential to enrich our wellbeing, fulfilling our innate psychological needs for competence and satisfaction (Kasser, 2017). Consider the simple act of reducing energy consumption, which may seem trivial but resonates deeply with our desire for mastery. This journey of exploration equips us with the knowledge and skills to navigate the complexities of energy conservation (Capstick et al., 2022).

The motivations behind pro-environmental actions are usually diverse. Some are driven by a noble impulse to contribute to the planet's wellbeing (Shin et al., 2022). Others, however, are guided by a more self-serving agenda, finding personal satisfaction in their environmental efforts (de Dominicis et al., 2017). The "warm glow" theory, proposed by Isen (1970), suggests that the positive emotions derived from acts of altruism can be powerful catalysts for prosocial behaviour. This theory implies that individuals may advocate for climate protection not solely out of altruism but also for the sheer joy it brings to their spirits. Recent insights reveal the potential of combining these motivators, creating a strong drive for pro-environmental actions (Hartmann et al., 2017).

Taufik et al. (2015) discovered an insight into the physiological realm – demonstrating that green behaviours can evoke a literal warmth, as individuals engaged in environmentally friendly actions exhibited higher body temperatures compared to their counterparts who refrained. This finding aligns with the psychological phenomenon known as the "helper's high", wherein acts of benevolence trigger a cascade of positive emotions within the human spirit (van der Linden, 2015). These vignettes come together highlighting the profound potential of PEB to enhance personal fulfilment and satisfaction.

Evolutionary traits: intrinsic versus extrinsic motivations

A wealth of studies highlights the profound physiological and psychological benefits of engaging in acts of kindness and moral conduct (Kahana et al., 2013; Wang et al., 2020). These virtues, deeply rooted in our evolutionary heritage, resonate in traits such as empathy and compassion (Moll et al., 2006). Extending a helping hand not only evokes positive emotions but also bestows upon the benefactor multiple physical and psychological rewards.

However, amidst human benevolence lies a challenge of the climate change. In the pursuit of fostering PEB, the efficacy of conventional approaches centred solely on extrinsic motivations, such as economic incentives, might not prove effective in the long run. Such motivations, fleeting by nature, fail to sustainably influence PEB over time (van der Linden, 2015).

In contrast, intrinsic motivations that propel individuals to undertake PEB out of a genuine belief in its righteousness and the pursuit of personal fulfilment – hold the promise of enduring change (Ryan & Deci, 2000; Capstick et al., 2022). This paradigm resonates with the tenets of self-determination theory, indicating that behaviours born from intrinsic motivation nurture personal wellbeing far more effectively than their extrinsically driven counterparts.

Embracing environmentally friendly behaviours not only shapes individuals' perceptions of themselves but also contributes to cultivating a more positive self-image, particularly when undertaken voluntarily rather than due to obligatory mandates such as legislation (Venhoeven et al., 2013, 2016). As we confront the daunting threat of environmental deterioration, the need for exploring intrinsic motivators in PEB becomes increasingly pressing. A growing body of evidence suggests that individuals harbour an innate inclination to contribute to a more sustainable and conserving society (van der Werff et al., 2013; Bopp et al., 2019; Venhoeven et al., 2020). Harnessing this intrinsic capacity for care towards others and the environment holds far greater promise than the fleeting nature of short-sighted incentives in persuading the public towards sustainable action (van der Linden, 2015; Venhoeven et al., 2020). It is imperative to delve deeper into the intricate dynamics of intrinsic motivations and their role in fostering PEB, as they hold the key to enduring change in our collective efforts to address pressing environmental challenges.

In exploring how psychological wellbeing relates to PEB, conventional wisdom suggests a trade-off between personal happiness and actions that support a sustainable environment. However, Brown and Kasser (2005) present a ground-breaking perspective that challenges this belief. They argue that individuals driven by intrinsic values and heightened self-awareness see psychological and environmental wellbeing not as conflicting but as inherently compatible.

Brown and Kasser's research demonstrates a shift in perspective, suggesting that prioritising personal wellbeing can encourage behaviours that contribute positively to the health of the natural environment. This departure from the traditional view of a trade-off between individual happiness and ecological responsibility offers a more nuanced understanding of how human wellbeing and environmental stewardship are interconnected.

Their research highlights the crucial role of intrinsic values – those stemming from core beliefs and authentic motivations. Individuals who embrace such values find harmony between their psychological wellbeing and their commitment to pro-environmental actions. Essentially, when people feel fulfilled, content, and aligned with their intrinsic values, they are more inclined to engage in behaviours that promote ecological health.

In essence, fostering psychological wellbeing benefits individuals and extends outward, positively influencing PEBs. This holistic perspective challenges the idea of a zero-sum game, suggesting that a flourishing psyche can coexist with an environmentally conscious lifestyle.

As society grapples with the urgent need for sustainable practices, these findings propose a fresh approach to fostering widespread environmental awareness. The research advocates for holistic strategies integrating mental health and ecological sustainability by emphasising the symbiotic relationship between personal wellbeing and PEBs. Encouraging individuals to nurture their psychological wellbeing becomes an investment in personal happiness and a strategic imperative for fostering a collective commitment to environmental preservation.

Brown and Kasser's pioneering work disrupts preconceived notions about the supposed conflict between personal happiness and PEBs. Their research lays the groundwork for a more integrative perspective, highlighting how intrinsic values, self-awareness, and psychological wellbeing align with active support for a healthy natural environment. This paradigm shift opens avenues for comprehensive strategies considering individual fulfilment and environmental stewardship as intertwined elements of a harmonious and sustainable future.

Self-determination theory

Discovering what truly drives human motivation and wellbeing has been a longstanding quest in psychology. In the 1980s, psychologists introduced Self-Determination Theory (SDT: Ryan & Deci, 2017), a comprehensive framework delving into the intricacies of human motivation and personality. Think of SDT as a macro-theory shedding light on the factors that either fuel or impede an individual's drive and overall wellbeing.

At the heart of SDT lie three fundamental psychological needs:

1 **Autonomy:** Our innate desire to feel in control, to exercise volition and choice in our actions. Autonomy is the driving force behind individuals seeking to align their behaviour with personal values and interests. In the realm of PEB, participating in environmental initiatives has the power to cultivate a sense of control. This sheds light on the challenges faced by individuals grappling with a sense of helplessness in the midst of the environmental crisis, making it hard for them to maintain sustained motivation.
2 **Competence:** This relates to our inherent need to experience effectiveness and mastery in navigating life's challenges. Competence is about feeling capable and confident, achieving desired outcomes and demonstrating an ability to influence the environment. This is a belief that we have the capability to alter the course of events. It hinges on the confidence in our skills and knowledge, empowering us to impact environmental outcomes.
3 **Relatedness:** The fundamental urge to connect with others, fostering a sense of belonging and attachment. This need highlights the significance of meaningful social connections, relationships, and the feeling of being cared for by others. In essence, it revolves around being part of a tribe that embraces our pro-environmental values, fostering a sense of being cared for. This can extend from our immediate community to the broader societal systems, encompassing entities like the government, economy, and legal system.

Therefore, motivations with a more internal locus of control tend to yield superior outcomes in terms of wellbeing, performance, and persistence. When individuals receive support in fulfilling their basic psychological needs, a ripple effect occurs, leading to heightened overall motivation and fulfilment.

Simultaneously, the last two and a half decades have seen a surge in positive psychological research delving into the intricacies of wellbeing. This line of inquiry acknowledges that wellbeing is a multifaceted concept shaped by individual values. For instance, in a study encompassing over 40,000 individuals across the EU, the researchers identified core elements of wellbeing that the majority of individuals need to experience for enhanced wellbeing (Huppert & So, 2009). These elements included the capacity for experiencing positive emotions like joy, happiness, hope, or awe; engagement, signifying individuals' full immersion and interest in their lives, enabling them to lose themselves in their passions; and a sense of meaning and purpose connected to something greater, providing a profound reason for our existence and an endeavour to contribute meaningfully to the world. Researchers recognise these elements as pivotal in ensuring overall wellbeing.

Simultaneously, it's crucial to consider a range of additional elements contributing to wellbeing. The significance of these elements can vary among individuals, and the absence of any doesn't necessarily label someone as unwell. In fact, individuals may still operate at peak levels, even if certain elements aren't fully cultivated. These supplementary factors encompass heightened levels of (1) self-esteem, (2) optimism, (3) resilience, (4) vitality, (5) self-determination, and (6) positive relationships. Individuals are encouraged to cultivate at least three of these six elements to enhance their quality of life. The choice of which elements to develop rests with individuals and aligns with the values they hold dear in their lives.

Moreover, the connection between PEBs and wellbeing isn't just a theory. Happier individuals are statistically more inclined to partake in actions benefiting society and the environment (Sulemana, 2016; Diener et al., 2018; Zhang et al., 2023). Nurturing the environment is linked to social wellbeing and plays a role in shaping a sense of community (Prati et al., 2016). It's not just a feel-good story; it amps up meaning in life, triggering a surge in positive vibes (Venhoeven et al., 2020). And it doesn't stop there; it's correlated with optimism, acting as a shield against anxiety and depression (Kaida & Kaida, 2019). So, despite the prevalent reports of eco-anxiety, insomnia, depression, and other challenges demonstrating concern for the environment and taking climate action can impact overall wellbeing.

Social norms and values

Individuals often find themselves entangled in a web of conflicting norms resonating profoundly within the context of PEB (McDonald et al., 2013). These conflicting social norms wield the power to either stifle or ignite action, laying bare the intricate influences that shape individuals' environmental choices.

PEB, often fraught with discomfort and inconvenience, is subject to diverse interpretations depending on an individual's norms, values, and the social circles they inhabit (Kahan, 2010). The approval or disdain of significant others holds sway over the perception of PEB; thus, their approval can increase or decrease our enthusiasm for engaging in PEB (Venhoeven et al., 2013).

Within the realm of these norms and values lies the complex and dynamic interplay of wellbeing (Venhoeven et al., 2013), which delve into hedonic (pleasure-based) and/or eudaimonic (purpose-based) wellbeing as contrasting yet intertwined concepts. While pursuing pleasure (hedonic) can engender virtuous action (eudaimonic) and vice versa, the relationship between PEB and wellbeing is nuanced. PEB may increase or reduce individual wellbeing, depending upon their convictions and beliefs.

Engagement in PEB might reduce hedonic wellbeing (pleasure) while fostering eudaimonic wellbeing, such as purpose. For effective PEB cultivation, eudaimonic (virtuous) values are imperative, enabling individuals to perceive these actions as morally upright and embrace them willingly (Venhoeven et al., 2013). This emphasis on intrinsic motivation assumes heightened significance in confronting enduring environmental crises like climate change, biodiversity loss, and ecological degradation, where sustained pro-environmental actions are imperative.

Beyond its palpable environmental ramifications, PEB provides a range of psychological, emotional, and social wellbeing benefits. A surge of recent research highlights the intricate nexus between eco-conscious actions and the broader spectrum of personal flourishing.

For example, a study centred on tourists showed that individuals experiencing higher levels of hedonic (pleasure-based) and eudaimonic (meaning-based) wellbeing exhibited a greater propensity towards PEBs (Lv et al., 2023). While refraining from dictating the direction of impact, the observed correlation hints at a potential interplay between wellbeing and PEB.

Similarly, our recent enquiry into beekeepers indicated not only that the wellbeing of farmers engaging in beekeeping increases, but this increase resonates across a range of established wellbeing theories, including the concept of psychological richness, wherein individuals undergo transformative growth and expansion (Burke & Corrigan, 2024). Thus, pro-environmental actions emerge as interventions that include a multitude of wellbeing theories, a discussion to be continued in the subsequent chapters.

Furthermore, our investigation expanded to include the wider beekeeping community, revealing a similar story. Bees alone trigger various motivations, prompting community members to engage in meaningful pro-environmental actions (Burke et al., under review). Participation in different aspects of community life leads to a notable increase in all aspects of wellbeing, from emotional satisfaction to improved psychological health. This positive change encompasses both the pursuit of happiness (hedonic wellbeing) and the pursuit of a meaningful life (eudaimonic wellbeing).

This evolving research challenges traditional viewpoints by suggesting that involvement in environmentally conscious activities goes beyond benefiting the planet and can contribute to an individual's overall wellbeing. As we navigate this intersection of environmentalism and wellbeing, it becomes increasingly clear that our actions don't just reflect our values; they shape our collective future.

What influences wellbeing in PEB

Imagine adopting a mindset that values actions geared towards the future, embodying the principle of delayed gratification. This perspective important in environmental sustainability, where individuals willingly sacrifice immediate gains in favour of a brighter tomorrow (House et al., 2004).

Conversely, those inclined towards short-term thinking prioritise instant satisfaction, often hesitating to make sacrifices as they focus on the present rather than considering the long-term consequences of their choices (House et al., 2004). Such individuals may also prioritise materialistic pursuits that conflict with environmental sustainability principles. The timeframe within which decisions are made emerges as a powerful force shaping environmental behaviours (Ashkanasy et al., 2004; Polonsky, 2011).

In sustainability, embracing a long-term perspective is intertwined with pro-environmental values and increased environmental engagement (Milfont et al., 2012). This viewpoint also influences how individuals perceive the future effects of their current actions and evaluate the outcomes of their decisions (House et al., 2004). Cultures prioritising a long-term orientation often show greater willingness to change behaviours, even at a perceived higher cost and requiring sacrifices (Chwialkowska et al., 2020).

Conventional approaches often emphasise external motivations such as taxation or financial incentives to influence PEB. Monetary incentives, like government subsidies for solar panels and electric vehicles, aim to address the social dilemma of cooperation. However, relying solely on cost-benefit analysis carries risks (Bolderdijk & Steg, 2014). Monetary rewards may unintentionally diminish intrinsic motivation for PEBs, promoting a mindset focused solely on financial calculations (Deci et al., 1999). This is difficult when cost-benefit analysis reinforces unsustainable behaviours, especially when financial stakes are relatively low. For example, achieving a 30% reduction in household energy consumption may result in marginal savings on utility bills, potentially insufficient to drive behavioural changes. Instead, framing energy conservation efforts in terms of CO_2 reduction may enhance perceived value and encourage more sustainable actions (Dogan et al., 2014).

Financial disincentives may have unintended consequences, leading individuals to rationalise behaviours that defy norms (Handgraaf et al., 2017). From a policy standpoint, disincentives may be seen as encroachments on personal autonomy, potentially sparking resistance (Schuitema et al., 2010). Importantly, individuals engaging in certain behaviours may not directly benefit financially from many types of PEB. For instance, renters, hotel guests, or employees may not directly pay for energy or water usage (Handgraaf et al.,

2017; Griffioen et al., 2019). When addressing habitual behaviours, relying solely on cost-benefit incentives derived from the analysis may lead to unintended consequences. Policy measures targeting intrinsic motivational factors could be crucial in overcoming these challenges.

Types of PEB that determine wellbeing

Engaging intimately with nature and its biodiversity transcends mere than solely appreciating its beauty; it emerges as a dynamic force, nurturing both personal wellbeing and environmental stewardship (Sandifer et al., 2015; Hausmann et al., 2016; Richardson & McEwan, 2018; Barragan-Jason et al., 2023). Through a systematic review of reviews exploring 832 distinct studies spanning 16 meta-analyses, Barragan-Jason and colleagues (2023) found a consistent narrative across diverse terrains, correlating nature immersion with conservationist tendencies and heightened states of wellness.

This interplay between humanity and nature is based on a reciprocity, wherein wellbeing and environmental guardianship reinforce one another. Nurturing these bonds between humans and nature demonstrates benefits for sustainability, surpassing conventional predictions (Barragan-Jason et al., 2023). This paradigm gains unprecedented potency when policies are written to resonate with individuals' intrinsic motivations, catalysing actions fuelled by the innate satisfaction derived from safeguarding the natural world.

Wellbeing offers policymakers a platform to delineate and evaluate the ramifications of diverse nature conservation strategies. For instance, probing into how the depletion of green spaces might impinge upon wellbeing becomes an important endeavour in managing sustainability (Diener et al., 2009). By embracing and harnessing the connections between human wellbeing and nature preservation, policymakers can develop initiatives that align with individuals' values, fostering a sustainable ethos of cohabitation with the environment.

Behaviour determinants

In the relentless pursuit of understanding the complexities surrounding the promotion of PEB, numerous reviews and meta-analyses were published, each aiming to analyse the effectiveness of various interventions. However, amidst these scholarly debates, clarity is needed, prompting a need for a more transparent understanding of which interventions have the most significant impact, under what conditions, and for what reasons. The factors guiding PEB are sometimes elusive.

Research focusing on Lithuania, a transitioning European Union nation, examined the years 2011–2020 (Minelgaitė & Liobikienė, 2021). This study, with its unique geographical and socio-economic context highlighted waste sorting and the adoption of environmentally friendly goods as leading behaviours experiencing a rapid increase, while water and energy conservation saw more modest growth.

These findings showed that the primary driver of PEB underwent a significant shift, with the perception of environmental issues taking precedence over self-transcendence values. Notably, self-enhancement values and awareness of behavioural consequences had a counterproductive effect, negatively influencing PEB. Analysing the fluctuation of these factors, researchers observed a minor change in self-enhancement values contrasted with significant transformations in other factors, particularly the rise in environmental responsibility.

As the global community, including us, grapples with the urgent need to reduce greenhouse gas emissions and address climate change, insights from this study highlight the importance of our role. They emphasise the importance of targeted interventions aligned with specific determinants. We are not just observers but active participants in this narrative with the power to make a difference.

On a broader scale, the effort to promote PEB resonates within the broader scholarly discussion. A ground-breaking classification system introduced by van Valkengoed and colleagues (2022) is a theoretical construct and a practical tool. It aims to decipher the puzzle surrounding intervention effectiveness, offering a structured understanding. This system connects six intervention types with 13 determinants of environmental behaviour, providing practitioners with a roadmap. It assists in selecting interventions tailored to the nuanced determinants guiding a particular behaviour, thus enhancing the likelihood of success within a given context. Essentially, it empowers us to navigate the complex landscape of behavioural change with strategic insight, fostering a more sustainable and environmentally aware world (van Valkengoed et al., 2022).

Individual or collective cultures

Embracing environmental stewardship entails navigating two distinct paths: the individual journey and the community endeavour. Rooted in cultural ethos, collectivism is a guiding principle that encourages individuals to find solace in shared goals and responsibilities (House et al., 2004). In tightly-knit collectivist societies like China or Brazil, social identity thrives on community, blending individual purpose with community wellbeing (Capstick et al., 2022). Conversely, personal pursuits often take precedence in individualistic landscapes like the UK or Denmark, even if their impacts extend to the broader community. Thus, pathways towards PEB diverge between individualistic and collectivist cultures, showcasing the significant influence of cultural values on environmental conduct.

This cultural backdrop acts as a lens through which societies view themselves and as a foundation for shaping sustainability outcomes. In collectivist societies, individual actions echo loudly within society, and the connections between people and their environment are tightly interwoven (Chwialkowska et al., 2020). Pro-environmental values and a strong commitment to sustainability flourish in these cultures (Liu & Segev, 2017). This commitment goes beyond mere lip service; individuals in collectivist societies demonstrate a remarkable personal commitment and willingly make sacrifices for the greater ecological good, whether through higher taxes or other contributions to environmental preservation (Owen & Videras, 2006). The collective conscience, linked to a long-term perspective, compels them towards a moral imperative, evoking a sense of remorse when the interests of future generations are neglected (Chwialkowska et al., 2020).

Moral obligation emerges as a pivotal force guiding individuals in collectivist cultures towards pro-environmental actions (van der Werff et al., 2013; Nguyen et al., 2016; Venhoeven et al., 2020). However, it's important to note that this sense of duty can also be a double-edged sword, as it may lead to guilt or resentment if individuals feel pressured to conform to societal expectations. It plays a central role in the decision-making arena of collectivist societies, where individuals navigate the complex interplay between personal and collective benefits (Chwialkowska et al., 2020). This profound sense of duty becomes a vital moderator, shaping the intricacies between pro-environmental intentions and actions and highlighting the potential for positive change when moral obligations are embraced.

In individualistic cultures, pro-environmental efforts often involve personal initiatives such as recycling or reducing energy consumption, contributing to individual wellbeing and satisfaction. For instance, in the UK, individuals may focus on reducing their carbon footprint through personal lifestyle choices, which can lead to a sense of personal accomplishment and wellbeing. Conversely, in collectivist cultures, the narrative of wellbeing intertwines with broader endeavours – envision community initiatives, collective movements, or advocacy for systemic improvements like robust public transportation. In China, for example, individuals may participate in community clean-up drives or support government-led initiatives for renewable energy, which can foster a sense of belonging and pride. Here, pro-environmental actions uplift individual wellbeing and enhance collective welfare, especially for those deeply connected to nature (Ibáñez-Rueda et al., 2020). Interestingly, Capstick et al. (2022) found limited evidence of a distinction in the relationship between PEBs and wellbeing in individualistic or collectivist societies.

From a policy perspective, it becomes clear that we must encompass both individual and collective efforts to amplify PEBs. This holistic approach recognises the synergies between personal and communal endeavours, in the pursuit of sustainable living. Driven by cultural undercurrents, policymakers would benefit from focusing on both the individual and the collective, fostering a shared commitment to a greener future. This emphasis on a holistic approach instills hope for effective change, acknowledging the power of collective action and individual responsibility in shaping a sustainable future. With its potential to facilitate communication, education, and behaviour change, technology can play a significant role in this process by providing tools and platforms for individuals and communities to engage in pro-environmental actions.

Socio-demographics of PEBs and wellbeing

When examining the intricate relationship between PEB and wellbeing, which encompasses physical, mental, and social health, various socio-demographic factors come to light, each leaving its distinct mark on the narrative. Let's embark on a journey to unravel some of these influences.

Age reveals a spectrum of evolving connections with nature, creating a dynamic portrait over time. Adolescence is a crucial period where peers plays a central role in shaping environmental attitudes (Collado et al., 2017). The subtle decline in PEBs as individuals transition through different life stages is a pressing issue. Understanding this trend becomes essential, providing insights for potential interventions and informed policies (Burke et al., 2023a).

Digging deeper, gender dynamics add complexity to the realm of PEB, significantly impacting subsequent wellbeing. Masculinity, often associated with dominance over nature, reflects a narrative of exploitation, overshadowing sustainability concerns (Hofstede, 2001). Conversely, femininity, with its collaborative relationship with nature, promotes stewardship over dominion. Aligned with these values, women emerge as inspiring champions of this harmonious connection, displaying a stronger bond with nature (Hughes et al., 2019).

Moreover, socio-economic status invites us to decipher its intricate interaction with PEB and wellbeing. Surprisingly, Capstick et al. (2022) uncovered a narrative suggesting that income or a country's development status may not solely dictate this relationship. Instead, a nuanced interplay between personal beliefs and societal expectations within socio-economic groups prompts us to unravel the layers of individual and social norms guiding our commitment to pro-environmental actions (Chan, 2020).

Exploring the socio-demographic influences on PEB and wellbeing is multifaceted, with age, gender, and socio-economic influencing human-environment interactions. Each aspect offers insights into the complexities of human behaviour, guiding our understanding and shaping our collective journey towards environmental stewardship.

Understanding the interplay between socio-economic status, PEB, and wellbeing is essential for developing effective strategies to promote environmental sustainability and improve wellbeing outcomes.

Firstly, Income level is a significant predictor of engagement in PEB. Recent research suggests that higher-income individuals are more likely to engage in environmentally friendly behaviours, such as purchasing eco-friendly products or investing in renewable energy (Thomas & Sharp, 2013). Higher-income levels give individuals greater resources and purchasing power to adopt sustainable practices. Additionally, higher income is associated with better overall wellbeing (Ng & Diener, 2019), which may positively influence attitudes and behaviours towards the environment.

Secondly, educational attainment is another important socio-economic factor influencing both PEB and wellbeing. It's not just a degree but a key that unlocks a world of environmental awareness and concern. Studies have shown that higher levels of education are associated with greater environmental awareness and concern (e.g., Gifford & Nilsson, 2014). Individuals with higher levels of education are more likely to have access to environmental information and resources, enabling them to make informed decisions and take pro-environmental actions (Milfont & Schultz, 2016). Furthermore, higher educational attainment is linked to better overall wellbeing outcomes, including higher life satisfaction and better mental health (Diener et al., 2010).

Thirdly, employment status can impact individuals' ability to engage in PEB and their overall wellbeing. The stability of employment is not just a source of income but a foundation for wellbeing. Research has shown that employed individuals may have more opportunities to participate in workplace sustainability initiatives or be able to afford environmentally friendly products and services (Bissing-Olson et al., 2013). Unemployment or precarious employment, on the other hand, may hinder individuals' ability to engage in PEB due to financial constraints or limited access to resources. Employment status is also closely linked to wellbeing, with stable employment providing individuals a sense of security and purpose (McKee-Ryan et al., 2005). It's crucial to address these issues to ensure the wellbeing of our society and the environment.

Disparities in access to resources and opportunities based on socio-economic status can influence individuals' ability to engage in PEB and their overall wellbeing. However, it's important to note that these disparities are not insurmountable. Individuals from lower socio-economic backgrounds may face barriers such as limited access to green spaces, public transportation, or affordable, sustainable options, which can hinder their ability to adopt environmentally friendly behaviours (Thomas & Sharp, 2013). Addressing socio-economic inequalities and improving access to resources and opportunities can pave the way for a more sustainable and equitable future, promoting environmental sustainability and wellbeing across diverse socio-economic groups.

Encouraging pro-environmental actions for wellbeing

The relationship between environmental awareness, the narrative surrounding PEB, and its impact on wellbeing is complex. For instance, while public consciousness of climate change and environmental issues appears robust, a paradox arises where environmental concern, while noble, can sometimes overshadow personal wellbeing. This can be seen in cases where individuals, driven by their environmental concerns, neglect their own mental health or social relationships.

The path of environmental stewardship often intersects with distressing emotions such as sleep disturbances, depression, and anxiety. However, the storyline gains complexity with the recognition of education and knowledge's significant role in shaping PEBs. *Knowledge*, in this context, refers to a comprehensive understanding of environmental issues, their causes, and potential solutions. It is a guiding compass, fuelling motivation and breaking down psychological barriers by empowering individuals with the tools and information they need to make a difference.

However, there is conflicting evidence when centred around enhancing eudaimonic wellbeing (e.g. life purpose) and intrinsic motivations. Here, the narrative shifts to a more empowering perspective, framing environmental actions as a journey towards personal fulfilment. This transformative factor not only benefits the environment but also enriches our lives, inspiring a sense of purpose and motivation.

Delving deeper, the supporting role of social capital emerges as a potent force. It's not merely about saving the planet but finding meaning and nurturing wellbeing. This narrative highlights the importance of individuals, each of us, becoming agents of positive change within institutions. We have the power to catalyse transformative shifts on various scales, instilling a sense of empowerment and responsibility.

Policy interventions become pivotal players in this unfolding narrative as the world grapples with environmental and societal challenges. However, public policies often overlook the intricate interplay between wellbeing and PEB. It is time for policymakers to take centre stage, understanding the nuanced evolution of PEB and aligning policies accordingly to foster these behaviours while conserving and restoring natural environments.

This demands a shift in focus. Instead of relying solely on external incentives like taxes and grants, policies should tap into intrinsic motivations for sustainability. The emphasis shifts towards nurturing voluntary PEB that emphasises the moral essence of sustainable action, making it inherently meaningful. The spotlight also turns to societal, collective actions, highlighting the benefits of communal sustainable endeavours and moving away from the exclusive focus on individual values and external rewards.

Amid this narrative, evidence suggests that females generally exhibit stronger PEBs than males. This observation is not to imply that all females are more environmentally conscious than all males but rather to highlight a trend that various social, cultural, and psychological factors may influence. Additionally, the evolving human-nature connections from childhood to adolescence and beyond create a dynamic backdrop with profound implications for human wellbeing and long-term environmental sustainability.

Importantly, one thing is abundantly clear: a one-size-fits-all approach to policy interventions may need re-evaluation in fostering widespread PEB. The narrative calls for targeted measures tailored to specific socio-demographic cohorts. This is not a luxury but a necessity. Recognising the diverse factors influencing wellbeing and environmental sustainability is crucial. Ignoring these nuances risks disrupting either our wellbeing and/or the planet's welfare, highlighting the need for immediate action.

Conclusion

A harsh reality becomes apparent in the complex web of our globalised economic system: it clashes with our environmental, social, and wellbeing priorities. This disharmony echoes through environmental degradation, economically manifested in the escalating costs of resource extraction, and socially through mounting tensions and deteriorating health indicators (McManus, 2023). Once hailed as a measure of prosperity, GDP is accused of driving biodiversity loss, pollution-related deaths, and worsening climate change. The relentless pursuit of economic growth has led to unsustainability encompassing environmental, economic, and social dimensions, demanding immediate attention.

Recognising this discord, a growing movement is veering away from the GDP-centric approach, advocating for broader wellbeing metrics that promise superior economic, social, and environmental outcomes. This chapter delved into the intricate relationship between PEB and wellbeing, exploring questions such as why environmental consciousness enhances wellbeing, which environmental actions contribute positively to wellbeing, and how socio-demographic factors intersect with PEB and wellbeing. Moreover, it explored strategies to encourage greater adoption of PEBs, promising a path to enhanced wellbeing and a sustainable future.

The narrative highlighted collective action as pivotal in tackling global environmental challenges (Masson & Fritsche, 2021). It explored the significance of intangible aspects like personal relationships and intrinsic motivations in promoting eudaimonic wellbeing, which lays the groundwork for enduring environmental sustainability. Shifting towards intrinsic values and motivations implies a collective moral duty to drive transformative change, crucial for advancing social and environmental welfare.

The COVID-19 pandemic is a stark reminder of our collective moral duty to safeguard societal wellbeing. Contrasted with environmental crises like climate change and biodiversity loss, it highlights the magnitude of threats to collective welfare. Lessons from the pandemic inform a blueprint for policy interventions emphasising collective moral obligations during adversity. There is a call for policies matching or surpassing the global ambition witnessed in the COVID-19 response to enhance societal wellbeing and urgently address environmental challenges. The narrative suggests that a profound shift towards policies centring on collective moral responsibilities is the key to a brighter future in the coming years.

References

Ashkanasy, N. M., Gupta, V., Mayfield, M. S., & Trevor-Roberts, E. (2004). *Future orientation* (pp. 282–342). Sage Publications. https://espace.library.uq.edu.au/view/UQ:70140

Barragan-Jason, G., Loreau, M., de Mazancourt, C., Singer, M. C., & Parmesan, C. (2023). Psychological and physical connections with nature improve both human well-being and nature conservation: A systematic review of meta-analyses. *Biological Conservation*, 277, N.PAG–N.PAG. Academic Search Complete.

Binder, M., Blankenberg, A.-K., & Guardiola, J. (2020). Does it have to be a sacrifice? Different notions of the good life, pro-environmental behavior and their heterogeneous impact on well-being. *Ecological Economics*, 167, N.PAG–N.PAG. Academic Search Complete.

Bissing-Olson, M. J., Iyer, A., Fielding, K. S., & Zacher, H. (2013). Relationships between daily affect and pro-environmental behavior at work: The moderating role of pro-environmental attitude. *Journal of Organizational Behavior (John Wiley & Sons, Inc.)*, 34(2), 156–175. https://doi.org/10.1002/job.1788

Bolderdijk, J., & Steg, L. (2014). *Promoting sustainable consumption: The risks of using financial incentives* (pp. 328–341). https://doi.org/10.4337/9781783471270.00033

Bopp, C., Engler, A., Poortvliet, P. M., & Jara-Rojas, R. (2019). The role of farmers' intrinsic motivation in the effectiveness of policy incentives to promote sustainable agricultural practices. *Journal of Environmental Management*, 244, 320–327. https://doi.org/10.1016/j.jenvman.2019.04.107

Brown, K. W., & Kasser, T. (2005). Are psychological and ecological well-being compatible? The role of values, mindfulness, and lifestyle. *Social Indicators Research*, 74, 349–368. https://doi.org/10.1007/s11205-004-8207-8

Burke, J., Clarke, D., & O'Keeffe, J. (2023a). Greening the mind: The power of integrating positive and environmental education for improving wellbeing. In G. Arslan & M. Yıldırım (Eds.), *Handbook of positive school psychology interventions: Evidence-based practice for promoting youth mental health* (pp. 145–159). Springer.

Burke, J., & Corrigan, S. (2024). Bee Well: A positive psychological impact of a pro-environmental intervention on beekeepers' and their families' wellbeing. *Frontiers in Psychology (Environmental Psychology)*, 15. https://doi.org/10.3389/fpsyg.2024.1354408

C40 Cities. (2019). *Measuring progress in Urban climate adaptation: Monitoring – Evaluating – Reporting framework*. Ramboll Fonden https://www.c40-knowledgehub.org/s/article/Measuring-Progress-in-Urban-Climate-Change-Adaptation-A-monitoring-evaluating-and-reporting-framework?language=en_US

Capstick, S., Nash, N., Whitmarsh, L., Poortinga, W., Haggar, P., & Brügger, A. (2022). The connection between subjective wellbeing and pro-environmental behaviour: Individual and cross-national characteristics in a seven-country study. *Environmental Science & Policy*, 133, 63–73. https://doi.org/10.1016/j.envsci.2022.02.025

Chan, H.-W. (2020). When do values promote pro-environmental behaviors? Multilevel evidence on the self-expression hypothesis. *Journal of Environmental Psychology*, 71, 101361. https://doi.org/10.1016/j.jenvp.2019.101361

Chwialkowska, A., Bhatti, W. A., & Glowik, M. (2020). The influence of cultural values on pro-environmental behavior. *Journal of Cleaner Production*, 268, 122305. https://doi.org/10.1016/j.jclepro.2020.122305

Collado, S., Evans, G. W., & Sorrel, M. A. (2017). The role of parents and best friends in children's pro-environmentalism: Differences according to age and gender. *Journal of Environmental Psychology*, 54, 27–37. https://doi.org/10.1016/j.jenvp.2017.09.007

Deci, E. L., Koestner, R., & Ryan, R. M. (1999). A meta-analytic review of experiments examining the effects of extrinsic rewards on intrinsic motivation. *Psychological Bulletin*, 125(6), 627–668; discussion 692–700. https://doi.org/10.1037/0033-2909.125.6.627

De Dominicis, S., Schultz, P. W., & Bonaiuto, M. (2017). Protecting the environment for self-interested reasons: Altruism is not the only pathway to sustainability. *Frontiers in Psychology*, 8, 1065. https://doi.org/10.3389/fpsyg.2017.01065.

Diener, E., Lucas, R., Schimmack, U., & Helliwell, J. (Eds.). (2009). *Well-being for public policy*. Oxford University Press. https://doi.org/10.1093/acprof:oso/9780195334074.002.0004

Diener, E., Oishi, S., & Tay, L. (2018). Advances in subjective well-being research. *Nature Human Behaviour*, 2(4), 253–260. https://doi.org/10.1038/s41562-018-0307-6

Diener, E., Wirtz, D., Tov, W., et al. (2010). New well-being measures: Short scales to assess flourishing and positive and negative feelings. *Social Indicatos Research*, 97, 143–156. https://doi.org/10.1007/s11205-009-9493-y

Dogan, E., Bolderdijk, J. W., & Steg, L. (2014). Making small numbers count: Environmental and financial feedback in promoting eco-driving behaviours. *Journal of Consumer Policy*, 37(3), 413–422. https://doi.org/10.1007/s10603-014-9259-z

European Commission (2021). *Delivering the European green deal*. https://commission.europa.eu/strategy-and-policy/priorities-2019-2024/european-green-deal/delivering-european-green-deal_en

Gifford, R., & Nilsson, A. (2014). Personal and social factors that influence pro-environmental concern and behaviour: A review. *International Journal of Psychology*, 49(3), 141–157. https://doi.org/10.1002/ijop.12034

Gong, Y., Li, Y., & Sun, Y. (2023). Waste sorting behaviors promote subjective well-being: A perspective of the self-nature association. *Waste Management*, 157, 249–255. Academic Search Complete.

Government of Iceland. (2011). *Iceland 2020 – Governmental policy statement for the economy and community knowledge, sustainability, welfare*. Prime Minister's Office. https://www.government.is/media/forsaetisraduneyti-media/media/2020/iceland2020.pdf

Government of Ireland. (2020). *Budget 2021: Wellbeing and the measurement of broader living standards in Ireland*. Government of Ireland. https://assets.gov.ie/90764/74a122af-0acf-4384-86b5-a0dbd6cca8f5.pdf

Griffioen, A. M., Handgraaf, M. J. J., & Antonides, G. (2019). Which construal level combinations generate the most effective interventions? A field experiment on energy conservation. *PloS One*, 14(1), e0209469. https://doi.org/10.1371/journal.pone.0209469

Handgraaf, M., Griffioen, A., Bolderdijk, J. W., & Thøgersen, J. (2017). Economic psychology and pro-environmental behaviour. In R. Ranyard (Ed.), *Economic psychology* (pp. 435–450). John Wiley & Sons, Ltd. https://doi.org/10.1002/9781118926352.ch27

Hartmann, P., Eisend, M., Apaolaza, V., & D'Souza, C. (2017). Warm glow vs. altruistic values: How important is intrinsic emotional reward in proenvironmental behavior? *Journal of Environmental Psychology*, 52, pp. 43–55. https://doi.org/10.1016/j.jenvp.2017.05.006

Hausmann, A., Slotow, R., Burns, J. K., & Minin, E. D. (2016). The ecosystem service of sense of place: Benefits for human well-being and biodiversity conservation. *Environmental Conservation*, 43(2), 117–127. https://doi.org/10.1017/S0376892915000314

Hickel, J. (2020). Quantifying national responsibility for climate breakdown: an equality-based attribution approach for carbon dioxide emissions in excess of the planetary boundary. *The Lancet. Planetary Health*, 4(9), e399–e404. https://doi.org/10.1016/S2542-5196(20)30196-0

Hofstede, G. (2001). Culture's recent consequences: Using dimension scores in theory and research. *International Journal of Cross Cultural Management*, 1(1), 11–17. https://doi.org/10.1177/147059580111002

House, R. J., Hanges, P. J., Javidan, M., Dorfman, P. W., & Gupta, V. (2004). *Culture, Leadership, and Organizations: The GLOBE Study of 62 Societies*. SAGE Publications.

Hughes, J., Rogerson, M., Barton, J., & Bragg, R. (2019). Age and connection to nature: When is engagement critical? *Frontiers in Ecology and the Environment*, 17(5), 265–269. https://doi.org/10.1002/fee.2035

Huppert, F. A., & So, T. T. C. (2009). What percentage of people in Europe are flourishing and what characterises them? *Well-being*, 1(1), 83–106.

Ibáñez-Rueda, N., Guillén-Royo, M., & Guardiola, J. (2020). Pro-environmental behavior, connectedness to nature, and wellbeing dimensions among granada students. *Sustainability*, 12(21), 9171.

IPBES. (2019). *Global assessment report on biodiversity and ecosystem services of the Intergovernmental Science-Policy Platform on Biodiversity and Ecosystem Services* (p. 1148). IPBES secretariat. https://doi.org/10.5281/zenodo.3553579

IPCC. (2022). *Climate Change 2022: Impacts, adaptation and vulnerability* (H.-O. Pörtner, D. C. Roberts, M. Tignor, E. S. Poloczanska, K. Mintenbeck, A. Alegría, M. Craig, S. Landgsdorf, S. Löschke, V. Möller, A. Okem, & B. Rama, Eds.). Cambridge University Press. https://www.ipcc.ch/report/ar6/wg2/

Isen, A. M. (1970). Success, failure, attention, and reaction to others: The warm glow of success. *Journal of Personality and Social Psychology*, 15(4), 294–301.

Kahan, D. (2010). Fixing the communications failure. *Nature*, 463(7279), Article 7279. https://doi.org/10.1038/463296a

Kahana, E., Bhatta, T., Lovegreen, L. D., Kahana, B., & Midlarsky, E. (2013). Altruism, helping, and volunteering: Pathways to well-being in late life. *Journal of Aging and Health*, 25(1), 159–187.

Kaida, N., & Kaida, K. (2019). Positive associations of optimism-pessimism orientation with pro-environmental behavior and subjective well-being: A longitudinal study on quality of life and everyday behavior. *Quality of Life Research*, 28(12), 3323–3332. Academic Search Complete.

Kasser, T. (2017). Living both well and sustainably: A review of the literature, with some reflections on future research, interventions and policy. *Philosophical Transactions of the Royal Society A: Mathematical, Physical and Engineering Sciences*, 375(2095), 20160369. https://doi.org/10.1098/rsta.2016.0369

Krasny, M. E., & Delia, J. (2015). Natural area stewardship as part of campus sustainability. *Journal of Cleaner Production*, 106, 87–96. Academic Search Complete.

Liu, Y., & Segev, S. (2017). Cultural orientations and environmental sustainability in households: A comparative analysis of Hispanics and non-Hispanic Whites in the United States. *International Journal of Consumer Studies*, 41(6), 587–596. https://doi.org/10.1111/ijcs.12370

Lv, X., Shi, K., Xu, S., & Wu, A. (2023). Will tourists' pro-environmental behavior influence their well-being? An examination from the perspective of warm glow theory. *Journal of Sustainable Tourism*, 1–18. https://doi.org/10.1080/09669582.2023.2243058

Masson, T., & Fritsche, I. (2021). We need climate change mitigation and climate change mitigation needs the 'We': A state-of-the-art review of social identity effects motivating climate change action. *Current Opinion in Behavioral Sciences*, 42, 89–96. https://doi.org/10.1016/j.cobeha.2021.04.006

McDonald, R. I., Fielding, K. S., & Louis, W. R. (2013). Energizing and de-motivating effects of norm-conflict. *Personality and Social Psychology Bulletin*, 39(1), 57–72. https://doi.org/10.1177/0146167212464234

McKee-Ryan, F., Song, Z., Wanberg, C. R., & Kinicki, A. J. (2005). Psychological and physical well-being during unemployment: A meta-analytic study. *Journal of Applied Psychology*, 90(1), 53–76. https://doi.org/10.1037/0021-9010.90.1.53

McManus, R. (2023). Exploring health and well-being in a European context. In M. L. De Lázaro Torres & R. De Miguel González (Eds.), *Sustainable development goals in Europe: A geographical approach* (pp. 45–64). Springer International Publishing. https://doi.org/10.1007/978-3-031-21614-5_3

Milfont, T. L., & Schultz, P. W. (2016). Culture and the natural environment. *Current Opinion in Psychology*, 8, 194–199. https://doi.org/10.1016/j.copsyc.2015.09.009

Milfont, T. L., Wilson, J., & Diniz, P. (2012). Time perspective and environmental engagement: A meta-analysis. *International Journal of Psychology*, 47(5), 325–334. https://doi.org/10.1080/00207594.2011.647029

Miller, T. (1995). *How to want what you have*. Avon.

Minelgaitė, A., & Liobikienė, G. (2021). Changes in pro-environmental behaviour and its determinants during long-term period in a transition country as Lithuania. *Environment, Development and Sustainability*, 23(11), 16083–16099. https://doi.org/10.1007/s10668-021-01329-9

Moll, J., Krueger, F., Zahn, R., Pardini, M., de Oliveira-Souza, R., & Grafman, J. (2006). Human fronto-mesolimbic networks guide decisions about charitable donation. *Proceedings of the National Academy of Sciences of the United States of America*, 103(42), 15623–15628. https://doi.org/10.1073/pnas.0604475103

Ng, W., & Diener, E. (2019). Affluence and subjective well-being: Does income inequality moderate their associations? *Applied Research Quality Life*, 14, 155–170. https://doi.org/10.1007/s11482-017-9585-9

Nguyen, T. N., Lobo, A., Nguyen, H. L., Phan, T. T. H., & Cao, T. K. (2016). Determinants influencing conservation behaviour: Perceptions of Vietnamese consumers. *Journal of Consumer Behaviour*, 15(6), 560–570. https://doi.org/10.1002/cb.1594

Owen, A. L., & Videras, J. (2006). Civic cooperation, pro-environment attitudes, and behavioral intentions. *Ecological Economics*, 58(4), 814–829. https://doi.org/10.1016/j.ecolecon.2005.09.007

Polonsky, M.J. (2011) Transformative Green Marketing: Impediments and Opportunities. *Journal of Business Research*, 64, 1311–1319. http://dx.doi.org/10.1016/j.jbusres.2011.01.016

Prati, G., Albanesi, C., & Pietrantoni, L. (2017). Social well-being and pro-environmental behavior: A cross-lagged panel design. *Human Ecology Review*, 23(1), 123–139. https://search.informit.org/doi/10.3316/informit.293860608339524

Richardson, M., & McEwan, K. (2018). 30 days wild and the relationships between engagement with nature's beauty, nature connectedness and well-being. *Frontiers in Psychology*, 9. https://www.frontiersin.org/articles/10.3389/fpsyg.2018.01500

Ryan, R. M., & Deci, E. L. (2000). Self-determination theory and the facilitation of intrinsic motivation, social development, and well-being. *American Psychologist*, 55(1), 68–78.

Ryan, R. M., & Deci, E. L. (2017). *Self-determination theory: Basic psychological needs in motivation, development, and wellness*. Guilford Press.

Sandifer, P. A., Sutton-Grier, A. E., & Ward, B. P. (2015). Exploring connections among nature, biodiversity, ecosystem services, and human health and well-being: Opportunities to enhance health and biodiversity conservation. *Ecosystem Services*, 12, 1–15. https://doi.org/10.1016/j.ecoser.2014.12.007

Schuitema, G., Steg, L., & Rothengatter, J. A. (2010). The acceptability, personal outcome expectations, and expected effects of transport pricing policies. *Journal of Environmental Psychology*, 30(4), 587–593. https://doi.org/10.1016/j.jenvp.2010.05.002

Shin, S., van Riper, C. J., Stedman, R. C., & Suski, C. D. (2022). The value of eudaimonia for understanding relationships among values and pro-environmental behavior. *Journal of Environmental Psychology*, 80, 101778. https://doi.org/10.1016/j.jenvp.2022.101778.

Steg, L., & Vlek, C. (2009). Encouraging pro-environmental behaviour: An integrative review and research agenda. *Journal of Environmental Psychology*, 29(3), 309–317. https://doi.org/10.1016/j.jenvp.2008.10.004

Stiglitz, J., Sen, A., & Fitoussi, J. P. (2009). *Report by the commission on the measurement of economic performance and social progress. Commission on the Measurement of Economic Performance and Social Progress.*

Sulemana, I. (2016). Are happier people more willing to make income sacrifices to protect the environment? *Social Indicators Research*, 127(1), 447–467.

Taufik, D., Bolderdijk, J. W., & Steg, L. (2015). Acting green elicits a literal warm glow. *Nature Climate Change*, 5(1), 37–40.

Thomas, C. and Sharp, V. (2013) Understanding the normalisation of recycling behaviour and its implications for other pro-environmental behaviours: A review of social norms and recycling. *Resources, Conservation and Recycling*, 79, 11–20. https://doi.org/10.1016/j.resconrec.2013.04.010

Tian, H., & Liu, X. (2022). Pro-environmental behavior research: Theoretical progress and future directions. *International Journal of Environmental Research and Public Health*, 19(11), 6721. https://doi.org/10.3390/ijerph19116721

van der Linden, S. (2015). Intrinsic motivation and pro-environmental behaviour. *Nature Climate Change*, 5(7), Article 7. https://doi.org/10.1038/nclimate2669

van der Werff, E., Steg, L., & Keizer, K. (2013). It is a moral issue: The relationship between environmental self-identity, obligation-based intrinsic motivation and pro-environmental behaviour. *Global Environmental Change*, 23(5), 1258–1265. https://doi.org/10.1016/j.gloenvcha.2013.07.018

van Valkengoed, A. M., Abrahamse, W., & Steg, L. (2022). To select effective interventions for pro-environmental behaviour change, we need to consider determinants of behaviour. *Nature Human Behaviour*, 6, 1482–1492. https://doi.org/10.1038/s41562-022-01473-w

Venhoeven, L. A., Bolderdijk, J. W., & Steg, L. (2013). Explaining the paradox: How pro-environmental behaviour can both Thwart and Foster well-being. *Sustainability*, 5(4), Article 4. https://doi.org/10.3390/su5041372

Venhoeven, L. A., Bolderdijk, J. W., & Steg, L. (2016). Why acting environmentally-friendly feels good: Exploring the role of self-image. *Frontiers in Psychology*, 7. https://www.frontiersin.org/articles/10.3389/fpsyg.2016.01846

Venhoeven, L. A., Bolderdijk, J. W., & Steg, L. (2020). Why going green feels good. *Journal of Environmental Psychology*, 71, 101492. https://doi.org/10.1016/j.jenvp.2020.101492

Wan, Q., & Du, W. (2022). Social capital, environmental knowledge, and pro-environmental behavior. *International Journal of Environmental Research and Public Health*, 19(3), 1443. https://doi.org/10.3390/ijerph19031443

Wang, Y., Ge, J., Zhang, H., Wang, H., & Xie, X. (2020). Altruistic behaviors relieve physical pain. *Proceedings of the National Academy of Sciences*, 117(2), 950–958. https://doi.org/10.1073/pnas.1911861117

Zhang, M., Zhang, W., & Shi, Y. (2023). Are happier adolescents more willing to protect the environment? Empirical evidence from Programme for International Student Assessment 2018. *Frontiers in Psychology*, 14, 1157409. https://doi.org/10.3389/fpsyg.2023.1157409

3 Natural vs urban environments vis-à-vis health and wellbeing

The interplay between environmental factors and human health has gained increasing attention from researchers, policymakers, and public health advocates in recent decades. Among these factors, pollution, urban design, and access to nature have emerged as pivotal determinants of mental and physical health outcomes and psychological wellbeing.

The rapid urbanisation and industrialisation witnessed worldwide have led to an alarming surge in pollution levels in our air, water, and soil. From exhaust fumes to industrial emissions, pollutants infiltrate our environment, posing immediate and significant health risks to populations living in urban areas. Moreover, the design of urban spaces plays a critical role in shaping human behaviour and wellbeing. Infrastructure, transportation systems, and architectural layouts can promote healthy living or exacerbate stress and sedentary lifestyles.

Amidst the built-up and congested nature of modern cities, access to natural environments has become increasingly scarce for many individuals. Yet, research consistently demonstrates the profound impact of nature exposure on human health and wellbeing. From reducing stress levels to enhancing cognitive function, the therapeutic effects of nature have been well-documented across various demographic groups.

This chapter delves into the intricate relationship between pollution, urban design, access to nature, and their collective influence on mental health, physical health, and psychological wellbeing. Through an exploration of empirical evidence, theoretical frameworks, and case studies, we aim to elucidate the complex mechanisms through which environmental factors shape human health outcomes. Moreover, we underscore the urgent need for interdisciplinary approaches and policy interventions in fostering healthier and more sustainable environments for present and future generations, emphasising the importance of collective action.

Drawing inspiration from pioneering urban design initiatives in cities like Adelaide, Lagos, and Singapore, this chapter showcases the transformative potential of urban planning in alleviating mental health challenges exacerbated by urban living. By reimagining urban spaces and enhancing nature

DOI: 10.4324/9781003452676-4

accessibility, cities can become catalysts for mental wellbeing, fostering resilience and community cohesion in the face of urban stressors.

By highlighting the multifaceted connections between our surroundings and our health, this chapter endeavours to inform efforts to promote environmental stewardship, foster equitable access to nature, and safeguard the wellbeing of individuals and communities worldwide.

Introduction

Humanity appears to be in a state of ongoing emergencies, each of which threatens our very existence. These range from non-communicable disease including diabetes, heart disease, cancer to global pandemics and drug resistance (World Health Organization, 2022). Alarmingly, a significant portion of these afflictions are intricately intertwined with the degradation of our environment (Dhimal et al., 2021).

While a lot is known about the causes of how the environment impacts our physical health, an exciting and growing area of research in psychology is the role the natural environment can also have in influencing our mental health. The aim of this chapter is to dissect the impact of various environmental aspects on the spectrum of health and wellbeing in the context of growing urbanisation.

Pollution and mental health

Pollution describes unwanted, often dangerous material that is introduced into Earth's environment from human activity, which threatens human health and damages ecosystems (Landrigan et al., 2018). The physical health impacts of pollution have been well documented. It is the largest environmental cause of disease and premature death globally, and was responsible for approximately nine million premature deaths in 2015, accounting for 16% of deaths worldwide in that year (Landrigan et al., 2018). Pollution-related disease and subsequent mortality can come from many sources including household or outdoor air, chemical pollution, noise pollution, soil pollution, or water pollution.

In addition to the physical health impacts of pollution, there is mounting evidence that environmental pollution is also damaging our mental health. Such are the effects of pollution on our mental health, that research has found that environmental pollution in its various forms can impact our mental wellbeing in numerous ways, including links to autism, depression, schizophrenia, cognitive deficits, mood, psychosis, and anxiety amongst others (Table 3.1).

Air pollution is an urgent concern, profoundly impacting the health and wellbeing of populations worldwide. This problematic issue entails a complex amalgamation of pollutants, notably PM2.5 delicate particulate matter and environmental ultrafine particles (UFPs), stemming from diverse human-driven activities such as hydrocarbon or wood combustion,

Table 3.1 Pollutants and their impacts on mental health

Topic	*Polluting agents*	*Human-related example sources*	*Impacts on mental health*
Air pollution	$PM_{2.5}$ fine particulate matter	Combustion of hydrocarbons or wood, agricultural activities, and dust particles	Linked to autism and depression
	Environmental ultrafine particles (UFPs)	Combustion of hydrocarbons or wood, biomass burning (i.e., agricultural burning, forest fires, and waste disposal), vehicular traffic and industrial emissions, tyre wear and tear from car brakes	Associated with schizophrenia, attention-deficit disorder
Heavy metals	Lead (Pb)	Metal smelting, battery manufacturing, and disposal, lead paint	Influences risks of psychosis
	Mercury	Mining, coal combustion, waste incineration	Related to autism and neurodevelopmental delay
Ionising radiations		Radioactive materials and radiation sources such as X-rays in medical devices	Associated with schizophrenia, depression, cognitive deficits
Organophosphate (OPs) pesticides		Agricultural pesticides	Linked to anxiety and depression
Light pollution		Streetlights, outdoor lighting, advertisements, large buildings Urban areas	Affects circadian rhythm with resulting mood disorders
Noise pollution		Road/rail/air traffic, buildings, construction	Risk of stress-related disorders, psychosis, and cognitive disorders
Environmental catastrophes		Wildfires, hydrocarbon environmental leaks, contamination of forests and ecosystems	Can lead to acute stress-reactions with fear, anxiety, insomnia, and concerns for mental and physical health. In the long term, increased depression, anxiety, and stress (post-traumatic stress disorder) are evident in individuals and communities exposed

Source: Adapted from Ventriglio et al. (2021).

agricultural practices, industrial emissions, and vehicular traffic. Beyond its detrimental effects on physical health, air pollution significantly encroaches upon mental wellbeing. Extensive research has unveiled a nexus between exposure to airborne contaminants and a spectrum of mental health disorders, encompassing autism, depression, schizophrenia, and attention-deficit disorder.

In tandem with air pollution, the pervasion of heavy metals poses a formidable threat to both ecosystems and human health. Lead (Pb) and mercury, predominantly emitted through industrial operations like metal refining, battery production, and coal burning, emerge as potent neurotoxins capable of impeding cognitive development and exacerbating mental health conditions, including autism.

Moreover, ionising radiations emanating from sources such as radioactive materials and medical equipment have been implicated in adverse mental health outcomes, including schizophrenia, depression, and cognitive impairment.

The widespread use of organophosphatepesticides in agricultural contexts has raised concerns regarding their impact on mental health. Studies have suggested a correlation between exposure to these chemicals and heightened levels of anxiety and depression.

Expanding beyond chemical pollutants, the proliferation of light and noise pollution within urban landscapes poses additional challenges to mental wellbeing. Light pollution from sources like streetlights and advertising disrupts natural circadian rhythms, exacerbating mood disorders. Concurrently, noise pollution generated by transportation and construction activities heightens the risk of stress-related disorders, psychosis, and cognitive impairments.

Furthermore, environmental catastrophes, such as wildfires and hydrocarbon leaks, not only present immediate physical hazards but also cast profound shadows over mental health. Individuals and communities subjected to such disasters often grapple with acute stress reactions marked by fear, anxiety, and sleep disturbances. In the long term, these traumatic events can precipitate elevated levels of depression, anxiety, and post-traumatic stress disorder, underscoring the intricate interplay between environmental degradation and mental wellbeing.

The evidence highlights the profound impact that both natural and urban environments and the pervasive pollution issue have on our psychological wellbeing. This, therefore, necessitates a discussion on environmental design and its potential to mitigate these effects. Subsequently, it examines how green spaces and ecological conservation can serve as essential tools in promoting mental wellbeing. By harnessing the restorative power of nature and addressing the environmental stressors outlined, we can create spaces that support and enhance mental health.

Urban greening: Adelaide, Australia

Adelaide, Australia demonstrates the profound impact of urban design on mental wellbeing, emphasising that a well-planned city can significantly contribute to the mental health of its residents. The city's strategy on green spaces (Healthy Parks Healthy People South Australia 2021–2026), focuses on improving public green spaces such as parks and conservation areas to create a connection with nature. The aim is to support physical activity and social interactions, both of which are vital for psychological wellbeing. The importance of inclusivity and community engagement in urban greening of Adelaide that reflects diverse cultural backgrounds and encourages community participation aims to enhance feelings of belonging and identity, reducing feelings of isolation and stress (Government of South Australia, 2021).

As in many cities globally, urban sprawl and environmental degradation threaten the mental wellbeing of Adelaide's residents. The city has identified ways to address these issues, including maintaining and expanding green spaces, improving access to public amenities, and ensuring city life is conducive to physical and mental health through careful planning and design (Government of South Australia, 2021).

Adelaide's policy initiatives, including the Healthy Parks Healthy People framework and efforts to increase urban greenery, exemplify how urban design can directly impact residents' mental wellbeing by creating spaces that encourage physical activity, relaxation, and social connection. It aims to prioritise mental health in urban planning to not only enhance individual wellbeing but also contribute to the overall health of the community, making the city more liveable and resilient (Hansen et al., 2020).

Within the complexity of human existence, adversities of all kinds possess the capacity to disrupt our mental equilibrium. Whether they manifest as profound life challenges or mundane daily struggles – such as family discord, sleepless nights, the stress of commuting, or the absence of nature in our daily lives – each element shapes our psychological wellbeing (Chowdhury, 2019). Nature's profound impact on mental health has gained widespread recognition, with studies indicating the restorative influence of natural environments on our psyche. Individuals who engage with nature, whether by strolling through lush landscapes or simply immersing themselves in natural settings, tend to experience reduced mental distress compared to those confined to urban environments (Bratman et al., 2015). This highlights the importance of nature in our lives.

This profound connection with the natural world provides solace and symbolises a transformative pathway to healing and wellbeing. It offers insights into turning adversity into growth and enriches our lives in inspiring ways. Nature's embrace affords us opportunities for self-care and introspection, enabling us to cultivate connections with both our inner selves and the external world. This perspective resonates with Allport's Model of Human Caring (MHC), emphasising the importance of nurturing these connections as a crucial step towards transitioning from distress to flourishing.

Today, there is a growing interest in exploring the intricate interplay between humanity and nature, driven by the myriad of health and wellbeing benefits associated with such interactions (Soga & Gaston, 2016; Ives et al., 2017; Pritchard et al., 2020). Importantly, this exploration also holds the potential to address pressing sustainability challenges (Zylstra et al., 2014), including climate change and biodiversity loss. This represents a quest for understanding and healing within the dynamic relationship between human existence and the natural world, offering hope and possibilities for renewal amidst the complexities of modern life.

Natural versus urban environments

Understanding nature's profound impact on human wellbeing is not a recent discovery but wisdom that has been passed down through the ages. It finds its roots in the teachings of Hippocrates, the revered father of medicine, who first espoused this belief as early as 400 BC (Wear, 2008). This belief that nature holds intrinsic benefits for our wellbeing has persisted through the centuries, and in recent years, scientific inquiry has only deepened our understanding of these phenomena.

The impact of nature on our mental health is not limited to a specific type of environment. Many scientific studies have shown that natural settings, with their rich biodiversity, lush vegetation, flowing water, fertile soil, and majestic trees, can significantly improve our wellbeing (Nejade et al., 2022). These environments can take various forms, from untouched wilderness to carefully designed spaces that maximise the benefits of nature. For instance, rooftop gardens, vertical green walls, and urban parks are all innovative ways to integrate nature into urban landscapes (Mantler & Logan, 2015). This adaptability is particularly evident in urban landscapes, where integrating nature is a creative response to spatial limitations. Thus, exploring nature's gifts transcends mere scientific enquiry, venturing into artistry, design, and human creativity. It stands as a testament to humanity's enduring quest to cultivate harmonious connections with the natural world, even amid the constraints of urban life.

Psychologist Stephen Kaplan's pioneering work in positive psychology unveils nature's transformative power, standing as a cornerstone. Kaplan's studies shed light on the unique characteristics of natural environments, essential for relieving mental fatigue through a process known as "attention restoration". This process refers to the ability of natural environments to replenish our mental energy, similar to how a good night's sleep restores our physical energy. His discoveries reveal individuals' heavy cognitive burden in urban environments, contrasted with the rejuvenating capacity inherent in natural landscapes. The urban environment, with its constant noise, bustling traffic, and perpetual human presence, perpetuates a state of cognitive overload, demanding constant attention. In stark contrast, the tranquillity of natural settings provides a sanctuary for mental peace and rejuvenation (Kaplan, 1995).

The links between environment and cognition have the potential to induce what psychologist Mihaly Csikszentmihalyi terms a "flow state", a mental state of complete absorption in the current experience. In this state, individuals are focused, energised, and enjoy the activity process. They lose their sense of time and often report feeling a sense of fulfilment and even ecstasy. Within this immersive experience, individuals discover a sense of fulfilment, embracing life's abundance. However, accessibility to nature remains a privilege, particularly for the burgeoning urban populace, which now exceeds 50% of the global population. This disparity underscores the urgent need to ensure that urban landscapes, constrained by spatial limitations, provide avenues for communion with nature. Moreover, this imperative extends to rural areas, where human encroachment threatens natural ecosystems and biodiversity, emphasising the universal mandate to safeguard and nurture our natural legacy. We all have a role to play in this. Whether you're a city planner, a community leader, or a concerned citizen, let's work together to integrate nature into our urban environments and preserve our natural legacy.

Green spaces and mental health

The benefits of interacting with nature, including psychological healing, improved body image, and resilience, are well-documented (Lackey et al., 2021). Yet, the robustness of objective data on these impacts often falls short of subjective experiences, that is, we are often dependent on self-reported measures of wellbeing in such studies (Coventry et al., 2021).

The quality and type of natural environments – ranging from urban green spaces to forests – affect wellbeing outcomes differently (Nguyen et al., 2021). For instance, high-quality natural woodland has been found to boost positive emotions more significantly than poorly designed, managed forests (Coventry et al., 2021). Equally, environmental factors like air quality and noise levels play a significant role in enhancing wellbeing (Gascon et al., 2018).

Preferences for nature experiences can vary by age and personal connection to the environment, with younger people deriving more significant benefits from untamed outdoor experiences (e.g., wild paths) than older people (e.g., paved paths) (Nguyen et al., 2021). For most people however, being in nature is linked to higher parasympathetic nervous system activity and supports relaxation, calmness, and contentment (Richardson et al., 2016).

This nuanced relationship between nature and wellbeing highlights the value of considering both the ecological quality of natural spaces and how different cohorts experience the benefits these provide. Enhancing the quality of green spaces and recognising the diverse needs and preferences within communities can play a critical role in maximising the positive effects of nature on mental health and wellbeing for the greater good. Put simply, we need to ensure the environments we live, work, visit, and grow old in are beneficial to everyone's mental wellbeing.

The role of nature in shaping wellbeing

Biophilia or nature therapy

The hypothesis that humans are inherently drawn to nature has been described as *biophilia* (Wilson, 2007), and suggests human affection for plants and other living things. Simply put, human evolution has evolved to live and thrive in a natural environment (Grinde & Patil, 2009), and we tend to be happier in natural settings. This connection to nature is not just cultural or intellectual; it is a fundamental part of our biology and psychology.

Two major theories have emerged into the ways that spending time in nature might affect our mental wellbeing. First, Attention Restoration Theory (ART) suggests that the mental fatigue associated with modern life reduces our capacity for direct attention. However, spending time in nature can support us in overcoming this mental fatigue and restoring our attention capacity (Kaplan, 1995). And second, the Stress Reduction Theory (SRT) describes how spending time in nature can influence our feelings or emotions by activating our parasympathetic nervous system. There is evidence to suggest that stress reduction is faster when a person is in natural compared to non-natural environments, for example, urban areas (Ulrich, 1984; Ulrich et al., 1991). The beauty and aesthetics of being in and seeing nature is therefore fundamental to our mental health (Grinde & Patil, 2009).

Below are examples of ways in which we can build biophilia into our lives:

- **Bring Nature Indoors:** add houseplants to your living/workspace, natural ornaments, for example, seashells, dried flowers, pinecones, acorns, pieces of wood.
- **Biophilic Interior Design:** bring natural light, nature-inspired images/art, using natural materials into your home or office, for example, wooden furniture, stone.
- **Get Outdoors:** make a conscious effort to spend more time in nature. Walking, hiking or sitting beside a waterbody helps us to reconnect with nature.
- **Mindfulness in Nature:** practice mindfulness *in nature*. Take 10–15 minutes to go outside. Observe and appreciate the beauty of nature and engage all of your senses. This could include watching birds or listening to birdsong, growing flowers or wild fruits or forest bathing for example.

Enhancing mental health and environmental outcomes through environmental design

More than 50% of people globally now live in cities, a figure which is expected to increase to 70% by 2050 (Bratman et al., 2015). But is living in cities good or bad for our health? The answer to this question is complex and mixed (Lau & Morse, 2008). Some studies have shown there are distinct advantages to living in urban areas compared to rural areas for our health, including in how we evaluate our lives (life evaluation), and how much worry, sadness, and anger we experience (negative affect) (Burger et al., 2020). Some have attributed an urban advantage to higher per capita income for urban compared to rural areas, where higher living standards can increase our wellbeing happiness (Veenhoven & Berg, 2013).

For others, however, urban areas are places where our health and wellbeing are at risk compared to rural areas (Krefis et al., 2018). This includes environment-related risks from heat stress, air pollution, and excessive noise, which can result in higher morbidity and mortality rates (Krefis et al., 2018). Our mental health is also at risk in cities, particularly compared to rural areas (Gruebner et al., 2017). For instance, overcrowded living conditions, public spaces and transport systems can increase our anxiety by reducing our personal space and creating a loss of control of our environment. Noise pollution can overload our senses, lead to irritability and decrease our cognitive functioning. Air pollution can also be detrimental to our mental wellbeing and is linked to higher rates of depression, anxiety, and cognitive abilities because of feelings of hopelessness and distress. Lack of greenspaces in urban areas deprives us of connecting with nature, an innate part of human evolution. This can increase stress levels and impact our mental wellbeing (Hari, 2019).

The conflicting views on how mental wellbeing differs for those of us that live in rural versus urban areas suggest one thing. Where mental wellbeing outcomes exist, these advantages need to be actively created and maintained through policy interventions rather than assumed to happen by chance (Rydin et al., 2012; Krefis et al., 2018). Given the growing prevalence of urban living now and in the future, we now turn to this point.

Urban design for mental health

We know a lot about the impacts of good urban design on our physical health like reducing obesity or lung diseases. However, we know a lot less about how urban areas impact our mental wellbeing. This is a concern because, as we discussed, urban areas have a significant impact on our mental wellbeing. Mental wellbeing also impacts cities, both directly (e.g., care costs, disability, physical health problems, mortality) and indirectly (e.g., discrimination, marginalisation, employment, informal carers, homelessness), with mental illness costs alone amounting to 4% of global Gross Domestic Product (OECD, 2014).

There is already growing evidence of how urban design can support our mental wellbeing. This knowledge can support architects, urban planners, developers, policymakers, communities, health professionals and others to support mental wellbeing outcomes (Centre for Urban Design and Mental Health, no date). Table 3.2 provides an overview of the ways in which urban design can improve our mental health.

Many of these solutions will also help us manage the multiple environmental crises of the 21st century that we face, including declining biodiversity, climate change and water, noise, and air pollution. For instance, concerns are emerging of the mental health challenges associated with climate change, including mental health challenges of increasing temperatures, trauma from extreme events, loss of culture and livelihoods (IPCC, 2023). Nature-based responses (see Table 3.2) can support human wellbeing, and mitigate the

Table 3.2 Urban design principles for improving mental wellbeing

Design principle	*How it impacts mental health*
Green space/Access to nature	Promotes exercise, setting for social interaction, biophilia theory, stress reduction theory, attention restoration theory
Active space for exercise	Increases serotonin for improved mood, improves sleep, enhances self-esteem, relieves stress, provides opportunities for social interaction
Pro-social places for social interactions	Improves self-esteem, self-confidence, empathy, increases sense of community, reduces loneliness, anxiety and isolation, enhances cognitive function
Urban safety	Feeling safe in daily life is important to our mental wellbeing, feeling unsafe can increase stress/anxiety, feeling unsafe can decrease physical activity/social interactions
Sleep	Noise and light pollution more likely to lead to sleep disturbances, which can negatively impact emotions and cognition, poor sleep (e.g., insomnia) can worsen mental health and vice versa
Transport and connection	Moves people efficiently and provides opportunities for education, employment, housing, social interaction, access to nature, all of which improve mental health, active transport (e.g., bike lanes/walking infrastructure) can improve mental health through physical activity, access to nature, safety, car commuting is associated stress, anxiety, aggression, and poor sleep
Economic stress and affordability	People living in poorer neighbourhoods likely to have lower self-esteem and high levels of frustration/ hopelessness, higher crime rates, and drug and alcohol use leading to feelings of fear/anxiety for others
Air pollution	Links to depression through neuroinflammatory reactions, dementia, and schizophrenia risks

Source: Adapted from (Centre for Urban Design and Mental Health, no date).

negative effects of these environmental crises at the same time. The growing recognition of links between our mental health, climate change adaptation and nature conservation demonstrate that taking action on improving the environment should be viewed as an opportunity, not least in terms of improving our mental wellbeing, and not as a cost.

Urban design in one of the fastest growing global cities: Lagos, Nigeria

Lagos, as one of the world's most densely populated cities, faces unique challenges in urban planning that significantly impact its inhabitants' mental wellbeing. As one of the fastest-growing metropolitan areas globally, the city is grappling with a range of planning issues, including reliable public transportation, affordable housing, access to public spaces, and the protection of wetlands, all of which impact mental wellbeing (Akindejoye et al., 2021).

Political and socio-economic factors currently limit the effectiveness of urban planning in addressing mental health concerns, however. These challenges present opportunities for Lagos to become a leading global city and a model for integrating mental wellbeing into urban design. This includes embedding mental health into city development plans, ensuring the availability of safe and aesthetically pleasing environments, and enhancing access to essential services, such as housing and public transport. Achieving these ambitions requires multidisciplinary expertise, involving government, the private sector, and community stakeholders, to create urban spaces that support mental wellbeing in Lagos. If designed well, it has the potential to make the city a global leader in promoting mental health through thoughtful urban design (Akindejoye et al., 2021).

Designing urban landscapes for mental health: Singapore

A *Contemplative Landscape Model* (CLM) is a useful tool to assess mental wellbeing. The model comprises seven specific aspects against which urban green space and mental wellbeing can evaluated, including: (i) layers of the landscape; (ii) different landforms present (topography); (iii) vegetation; (iv) colour and light; (v) compatibility (harmony and balance between natural and created elements); (vi) archetypal elements (landscape elements loaded with significance, e.g., waterfall, trees), and; (vii) peace and silence.

Research in Singapore has shown that certain CLM features, particularly urban landscapes that provide peace and silence, comprise different layers, and have archetypal elements, significantly predict positive emotions and brain activity indicative of relaxation and mindfulness in nature.

This highlights the importance of specific landscape features in designing urban green spaces for mental health benefits, suggesting that features enabling visibility of landscape layers, presence of symbolic elements, and areas for peace and solitude can enhance wellbeing among city dwellers (Olszewska-Guizzo et al., 2022).

Drawing on the CLM, examples of strategies to include in urban design include:

- opening views to far-away scenery so that visitors can see both nearby and far-away objects (layers of the landscape);
- creating visual and noise buffers dividing public green spaces from the city environment, planning for comfortable seating for solitary resting (peace and silence);
- highlighting the existing archetypal element with design so that it dominates the view (e.g., by clearing the surrounding of a solitary tree to make its silhouette more distinct) (archetypal features);
- incorporating more naturalistic planting plans (seemingly planted by nature) including spontaneous and diverse vegetation that displays seasonal and diurnal changes (compatibility).

Urbanisation is one of the defining demographic trends of the modern era. As more of us live in and move to urban areas during the 21st century, there are impacts on our wellbeing that are only beginning to be understood and documented. This section has discussed how our mental wellbeing is particularly vulnerable to urbanisation but that low-risk/high-reward opportunities exist to make our urban spaces places that can improve our natural environments and our mental wellbeing (see Table 3.2 and case studies). It is imperative that we (architects, urban planners, developers, policymakers, individuals, communities, and health professionals) embrace the opportunities to improve both environmental and health outcomes in urban areas during the 21st century.

Despite the opportunities to improve our mental wellbeing by designing for nature and wellbeing, we are in the midst of a series of environmental crises (e.g., climate change, biodiversity loss, ocean acidification, deforestation, land degradation) that are already having a profound impact on people's mental wellbeing, often associated with a sense of loss, anger, and frustration at the deterioration in our natural world caused by humans.

The need for evidence

Developing nature-positive cities presents numerous challenges encompassing the design, funding, implementation, and maintenance of urban blue and green spaces. These hurdles range from bureaucratic obstacles to community resistance, each presenting its unique trial. Many difficulties arise from insufficient funds, political inertia, and community aversion to change. However, a significant barrier lies in the need for more evidence demonstrating how specific projects will benefit local communities, the city as a whole, and the budget of responsible government departments. Budget constraints often clash with ambitious visions, jeopardising project feasibility. While numerous examples of poorly designed and low-return projects exist, the government aims to ensure a return on investment or, at the very least, value for money. Incorporating nature into cities poses a challenge as evaluating the benefits of nature-based solutions proves challenging, especially compared to more traditional grey infrastructure. Nevertheless, when designed effectively, green infrastructure has the potential to yield multiple benefits simultaneously.

Effective stakeholder engagement is a significant and intricate challenge, yet it is crucial to ensure successful project outcomes. Community buy-in can heavily influence whether a project is implemented, how it is designed, and how it is maintained. When the community accepts and embraces a project, it is more likely to succeed, both in the short term during implementation and in the long term, when maintenance and utilisation are essential for realising benefits. However, coordinating stakeholders requires navigating a delicate balance, as it often involves overcoming communication barriers and managing conflicting interests. Addressing these concerns necessitates meticulous planning and striking a balance between progress and sustainability.

Every project presents unique circumstances. While some may focus on new sites where nature-based solutions can be integrated into the design, "green field sites" may not be available in urban contexts. Instead, progress may necessitate retrofitting or redeveloping existing infrastructure. Legacy infrastructure poses constraints, demanding innovative solutions to modernise and adapt often outdated systems. Cultural heritage preservation introduces additional layers of complexity, as it involves blending modernity with history. However, within these obstacles lies the potential for transformative change, propelling progress in urban landscapes.

The lack of evidence available to decision-makers is frequently cited as a significant barrier to creating new, high-quality blue and green city spaces. Initiatives like the Valuing Natural Capital in Communities for Health (VNiC-Health) in Ireland project aim to tackle this challenge by bridging environmental science and psychological wellbeing. While it's understood that spending time in nature offers numerous benefits, we are only beginning to grasp how these benefits manifest, how to measure them effectively, and most importantly, how to accurately evaluate the myriad advantages provided by our natural environment. The VNiC-Health Project endeavours to address these gaps by developing methodologies where key decision-makers, such as government officials, local councils, or practitioners, can engage with communities, gather crucial information, and assess the presence and design of blue and green spaces in urban areas. This comprehensive approach aims to justify funding and implementation decisions. Furthermore, this adaptable strategy will serve as a blueprint for integrating nature-based solutions and promoting health and wellbeing while simultaneously delivering a range of essential ecosystem services.

Conclusion

The intricate connections between pollution, urban design, and access to nature highlight the profound implications for human health and wellbeing.

As we navigate the challenges posed by rapid urbanisation, industrialisation, and environmental degradation, it becomes increasingly evident that our actions not only shape the health of our planet but also profoundly impact our physical and mental health.

The evidence presented in this chapter highlights the pivotal role of pro-environmental behaviours in mitigating the adverse effects of pollution and urban stressors on human health. By reducing carbon emissions, minimising waste, and advocating for sustainable urban planning, individuals and communities can contribute to cleaner air, water, and green spaces, thereby fostering healthier environments for all.

Furthermore, promoting equitable access to nature is essential for addressing health disparities and enhancing psychological wellbeing. By ensuring that green spaces are accessible to all members of society, regardless of socio-economic status or geographic location, we can harness the restorative power of nature to promote resilience, reduce stress, and improve overall quality of life.

Fostering a symbiotic relationship between humans and the environment is essential for building resilient, sustainable communities. By prioritising pro-environmental behaviours, advocating for greener urban spaces, and protecting natural ecosystems, we can create environments that not only support human health and wellbeing but also preserve our planet's biodiversity and ecological balance for future generations.

As we move forward, we must continue integrating environmental considerations into public health policies, urban planning initiatives, and community-based interventions. Recognising the interconnectedness of environmental, social, and health outcomes, we can chart a course towards a more sustainable and resilient future for all.

In the end, our planet's and its inhabitants' health are inextricably linked. By embracing pro-environmental behaviours and fostering environments that promote clean air, vibrant green spaces, and sustainable living, we can create a healthier, more equitable world for future generations.

References

Akindejoye, F., Ezedinma, U., & Ike, N. (2021). A case study of urban design for wellbeing and mental health in Lagos, Nigeria. *Journal of Urban Design and Mental Health*, 7(10).

Bratman, G. N., Hamilton, J. P., Hahn, K. S., & Gross, J. J. (2015). Nature experience reduces rumination and subgenual prefrontal cortex activation. *Proceedings of the National Academy of Sciences*, 112(28), 8567–8572. https://doi.org/10.1073/pnas.1510459112

Burger, M., Morrison, P. S., Hendriks, M., & Hoogerbrugge, M. M. (2020). *Urban-rural happiness differentials across the world*. UN Sustainable Development Solutions Network. https://worldhappiness.report/ed/2020/urban-rural-happiness-differentials-across-the-world/ (Accessed: 11 March 2024).

Centre for Urban Design and Mental Health. (no date). *How mental health affects the city*. Centre for Urban Design and Mental Health. https://www.urbandesignmentalhealth.com/how-mental-health-affects-the-city.html (Accessed: 11 March 2024).

Chowdhury, M. H. (2019). *What is the mental health continuum model?* PositivePsychology.com. https://positivepsychology.com/mental-health-continuum-model/ (Accessed: 28 February 2024).

Coventry, P. A., et al. (2021). Nature-based outdoor activities for mental and physical health: Systematic review and meta-analysis. *SSM - Population Health*, 16, 100934. https://doi.org/10.1016/j.ssmph.2021.100934

Dhimal, M., Neupane, T., & Lamichhane Dhimal, M. (2021). Understanding linkages between environmental risk factors and noncommunicable diseases-A review. *FASEB bioAdvances*, 3(5), 287–294. https://doi.org/10.1096/fba.2020-00119

Gascon, M., et al. (2018). Long-term exposure to residential green and blue spaces and anxiety and depression in adults: A cross-sectional study. *Environmental Research*, 162, 231–239. https://doi.org/10.1016/j.envres.2018.01.012

Government of South Australia. (2021). *Health parks healthy people South Australia 2021–2026*. Government of South Australia. https://cdn.environment.sa.gov.au/environment/docs/healthy_parks_healthy_people_2021_aprl.pdf

Grinde, B., & Patil, G. G. (2009). Biophilia: Does visual contact with nature impact on health and well-being? *International Journal of Environmental Research and Public Health*, 6(9), 2332–2343. https://doi.org/10.3390/ijerph6092332

Gruebner, O., Rapp, M. A., Adli, M., Kluge, U., Galea, S., & Heinz, A. (2017). Cities and mental health. *Deutsches Ärzteblatt International*, 114(8), 121–127. https://doi.org/10.3238/arztebl.2017.0121

Hansen, T., et al. (2020). A case study of urban design for wellbeing and mental health in Adelaide, Australia. *Journal of Urban Design and Mental Health*, 6(7). https://www.urbandesignmentalhealth.com/journal-6-adelaide.html

Hari, J. (2019). *Lost connections: Why you're depressed and how to find hope*. Bloomsbury Publishing Plc.

IPCC. (2023). *Climate Change 2023: Synthesis Report. Contribution of Working Groups I, II and III to the Sixth Assessment Report of the Intergovernmental Panel on Climate Change* (Core Writing Team, H. Lee, & J. Romero, Eds.). IPCC. https://doi.org/10.59327/IPCC/AR6-9789291691647

Ives, C. D., et al. (2017). Human–nature connection: A multidisciplinary review. *Current Opinion in Environmental Sustainability*, 26–27, 106–113. https://doi.org/10.1016/j.cosust.2017.05.005

Kaplan, S. (1995). The restorative benefits of nature: Toward an integrative framework. *Journal of Environmental Psychology*, 15(3), 169–182. https://doi.org/10.1016/0272-4944(95)90001-2

Krefis, A. C., Augustin, M., Schlünzen, K. H., Oßenbrügge, J., & Augustin, J. (2018). How does the urban environment affect health and well-being? A systematic review. *Urban Science*, 2(1), 21. https://doi.org/10.3390/urbansci2010021

Lackey, N. Q., Tysor, D. A., McNay, G. D., Joyner, L., Baker, K. H., & Hodge, C. (2021). Mental health benefits of nature-based recreation: A systematic review. *Annals of Leisure Research*, 24(3), 379–393. https://doi.org/10.1080/11745398.2019.1655459

Landrigan, P. J., et al. (2018). The Lancet Commission on pollution and health. *The Lancet*, 391(10119), 462–512. https://doi.org/10.1016/S0140-6736(17)32345-0

Lau, R., & Morse, C. A. (2008). Health and wellbeing of older people in Anglo-Australian and Italian-Australian communities: A rural–urban comparison. *Australian Journal of Rural Health*, 16(1), 5–11. https://doi.org/10.1111/j.1440-1584.2007.00933.x

Mantler, A., & Logan, A. C. (2015). Natural environments and mental health. *Advances in Integrative Medicine*, 2(1), 5–12. https://doi.org/10.1016/j.aimed.2015.03.002

Nejade, R. M., Grace, D., & Bowman, L. R. (2022). What is the impact of nature on human health? A scoping review of the literature. *Journal of Global Health*, 12, 04099. https://doi.org/10.7189/jogh.12.04099

Nguyen, P.-Y., et al. (2021). Green space quality and health: A systematic review. *International Journal of Environmental Research and Public Health*, 18(21), 11028. https://doi.org/10.3390/ijerph182111028

OECD. (2014). *Making mental health count: The social and economic costs of neglecting mental health care*. Organisation for Economic Co-operation and Development. https://www.oecd-ilibrary.org/social-issues-migration-health/making-mental-health-count_9789264208445-en (Accessed: 11 March 2024).

Olszewska-Guizzo, A., Sia, A., Fogel, A., & Ho, R. (2022). Features of urban green spaces associated with positive emotions, mindfulness and relaxation. *Scientific Reports*, 12(1), 20695. https://doi.org/10.1038/s41598-022-24637-0

Pritchard, A., Richardson, M., Sheffield, D., & McEwan, K. (2020). The relationship between nature connectedness and eudaimonic well-being: A meta-analysis. *Journal of Happiness Studies*, 21(3), 1145–1167. https://doi.org/10.1007/s10902-019-00118-6

Richardson, M., McEwan, K., Maratos, F., & Sheffield, D. (2016). Joy and calm: How an evolutionary functional model of affect regulation informs positive emotions in nature. *Evolutionary Psychological Science*, 2(4), 308–320. https://doi.org/10.1007/s40806-016-0065-5

Rydin, Y., et al. (2012). Shaping cities for health: Complexity and the planning of urban environments in the 21st century. *The Lancet*, 379(9831), 2079–2108. https://doi.org/10.1016/S0140-6736(12)60435-8

Soga, M., & Gaston, K. J. (2016). Extinction of experience: The loss of human–nature interactions. *Frontiers in Ecology and the Environment*, 14(2), 94–101. https://doi.org/10.1002/fee.1225

Ulrich, R. S. (1984). View through a window may influence recovery from surgery. *Science*, 224(4647), 420–421. https://doi.org/10.1126/science.6143402

Ulrich, R. S., Simons, R. F., Losito, B. D., Fiorito, E., Miles, M. A., & Zelson, M. (1991). Stress recovery during exposure to natural and urban environments. *Journal of Environmental Psychology*, 11(3), 201–230. https://doi.org/10.1016/S0272-4944(05)80184-7

Veenhoven, R., & Berg, M. (2013). Has modernisation gone too far? Modernity and happiness in 141 contemporary nations. *International Journal of Happiness and Development*, 1(2), 172. https://doi.org/10.1504/IJHD.2013.055645

Ventriglio, A., et al. (2021). Environmental pollution and mental health: A narrative review of literature. *CNS Spectrums*, 26(1), 51–61.

Wear, A. (2008). Place, health, and disease: The airs, waters, places tradition in early modern England and North America. *Journal of Medieval and Early Modern Studies*, 38(3), 443–465. https://doi.org/10.1215/10829636-2008-003

Wilson, E. O. (2007). Biophilia and the conservation ethic. In G. Holmqvist & C.Lundqvist-Persson (Eds.), *Evolutionary perspectives on environmental problems* (263–272). Routledge.

World Health Organization. (2022). *WHO top threats to global health - Special article collection, annual reviews*. https://www.annualreviews.org/page/worldhealththreats (Accessed: 15 March 2024).

Zylstra, M. J., Knight, A. T., Esler, K. J., & Le Grange, L. L. L. (2014). Connectedness as a core conservation concern: An interdisciplinary review of theory and a call for practice. *Springer Science Reviews*, 2(1), 119–143. https://doi.org/10.1007/s40362-014-0021-3

Part 2

Individuals' wellbeing, pro-environmental behaviour and biodiversity

In the forthcoming section of this book, Part 2, we embark on an exploration of the various perspectives on individuals' wellbeing. These perspectives are crucial to understand and serve as lenses through which we can comprehend the impact of pro-environmental actions on health and overall wellbeing of individuals.

In Chapter 4, we delve into the contrasting viewpoints of deficit-based and abundance-based approaches to wellbeing, contextualising them within the realm of health. Through an analysis of existing evidence, we clarify the current understanding of how pro-environmental behaviour, such as recycling or reducing carbon footprint, intersects with these perspectives.

Chapter 5 shifts the focus to the realm of positive psychology, specifically delving into the emotional aspects of wellbeing. Here, we examine a range of theories, some of which have already found application in environmental research, while others are still emerging. We offer practical examples of how insights from positive psychology, which you can readily apply, can be leveraged in environmental contexts to both assess and enhance human wellbeing.

Moving forward to Chapter 6, we explore the eudaimonic components of wellbeing using a range of positive psychological models. We investigate how these aspects relate to pro-environmental behaviour and the protection of biodiversity. By exploring these connections, we aim to deepen the understanding of the motivations behind sustainable actions. In the following chapters, we will discuss the role of social support and community engagement in pro-environmental behaviour and the potential of mindfulness and self-compassion in promoting sustainable actions.

Finally, in Chapter 7, we expand our exploration to encompass mindsets as guideposts for wellbeing. These are other concepts within positive psychology that can be readily adapted to promote human health and wellbeing. We emphasise that these concepts require choices to be made to impact wellbeing positively or negatively while attending to pro-environmental cause. This chapter delves into the flexibility of these concepts and their potential to influence positive change.

DOI: 10.4324/9781003452676-5

By the conclusion of Part 2, readers will have gained insights into a spectrum of positive psychology theories, models, and applications. Armed with this knowledge, we can better understand how to foster pro-environmental behaviour and contribute to the protection and enhancement of biodiversity.

4 Environment in the context of illbeing and wellbeing

When people think of biodiversity, their minds often turn to bees, flowers, and trees, yet it encompasses much more. The ocean presents another biodiverse haven for those fortunate enough to reside in coastal regions. Mary, an ocean educator, passionately imparts knowledge about the wonders of our oceans to people of all ages. Embracing her role, Mary conducts teaching sessions both in the classroom and by the seaside.

Mary finds the beachside sessions particularly captivating. Once the children step outside along the oceanfront, she observes a noticeable shift in their behaviour. Infused with curiosity and wonder, they eagerly explore rockpools and seaweed, delighting in discovering crabs and other sea creatures. Mary notes joyfully how the children immerse themselves in the fresh air and the excitement of exploration.

Yet, it's not only the children who exhibit this exuberance. During training sessions, the inner child within adults often emerges as they partake in activities like building sandcastles or sculpting sand. Mary acknowledges the unique allure of the ocean, recognising that it captivates people in different ways – be it through its sounds, smells, or the sensation of sand between their toes. Regardless of the specific allure, what matters, she emphasises, is the connection forged.

Mary also highlights the deep connection Indigenous peoples across the globe have with their environments. Over generations, they have amassed a wealth of knowledge about their habitats, often rooted in traditional wisdom rather than scientific understanding. Yet, Mary stresses, this knowledge is just as vital, if not more so, as it underscores the profound connection between people and their surroundings. Ultimately, she concludes that while education is essential, the connection is paramount.

In Co. Donegal, surf therapy offers one-on-one instruction tailored to youth with mental and physical disabilities, including Autism Spectrum Disorder, ensuring that each child's experience is customised to their needs and goals. This therapy has shown great potential benefits, particularly for children with autism, with evidence indicating significantly reduced anxiety and increased social and natural connectedness. Moreover, participants report positive outcomes regarding confidence, self-worth, and interpersonal skills.

DOI: 10.4324/9781003452676-6

Parents also benefit from social connections and peer support, finding a sense of respite and escape while at the beach (Britton 2020).

Continuing with ocean-related activities, beach clean-ups conducted by Clean Coasts at various locations around Ireland offer a high potential as a health and wellbeing intervention. These clean-ups can be implemented at any beach with minimal to no cost. Participants gather under the guidance of a team leader to clean the beach and remove rubbish. Marine litter threatens ocean biodiversity, with items like plastic bags often mistaken for jellyfish by turtles and microplastics found in the bodies of deceased marine life.

Engaging in beach clean-ups fosters a sense of inclusion and community building, with participants feeling they are taking proactive environmental action and assuming responsibility. Additionally, participants value the knowledge shared by group leaders and appreciate the opportunity to learn more about ocean literacy. Not to mention it creates a positive impact on everyone's wellbeing with many participants mentioning the health outcomes of being outside.

Initiatives, such as this one are very helpful amid emerging challenges which threaten our wellbeing, including a global mental health crisis. For instance, approximately 40% of adults have at one point in their lives experienced a mental health disorder and one in ten have attempted suicide (Hyland et al., 2022). These disorders often first emerge in childhood, adolescence, or young adulthood and can play out as significant mental health risks for the remainder of our life. Despite the improvements in our understanding of mental health over the last 50 years, the reality is that too many of us experience a mental health disorder. The causes of what is making one-in-two of us mentally unwell during our lifetime can include weak social supports, past traumas, family or financial worries, poor physical health, substance abuse, and many other factors besides (Tost et al., 2015). At the same time, our environment is under immense pressure and is being degraded at alarming rates, from everything to climate breakdown, biodiversity loss, land degradation, deforestation, ocean acidification, and pollution in its many forms (e.g., air, water, chemical) (Richardson et al., 2023). Both human and environmental wellbeing are connected.

Mental health and environment connection

As the world grapples with a multitude of crises, including the increase in non-communicable diseases, the threat of pandemics, and the deterioration of our environment, the connection between our surroundings and mental health has become a pressing issue. We emphasise the necessity of understanding how our environment influences our psychological state, advocating for a comprehensive mental health approach incorporating environmental factors.

In discussions about the effects of pro-environmental actions, anxiety and depression are frequently mentioned. However, recent findings suggest that biodiversity and engaging in such actions can also serve as sources of deeper meaning, engagement, positive emotions, self-efficacy, and other beneficial traits that contribute to long-term improvements in health for many individuals. This chapter aims to introduce a positive psychology perspective on the impact of pro-environmental actions on wellbeing. Positive psychology, with its focus on understanding the strengths and virtues that contribute to wellbeing rather than focusing solely on deficits, offers a pathway for personal growth and empowerment through pro-environmental actions. Recognising that negative feelings (e.g., anxiety) and positive states (e.g., meaning and purpose) can coexist, this chapter will explore how these phenomena intersect with pro-environmental action and mutually enhance psychological, emotional, and social wellbeing.

Additionally, the chapter will delve into a range of alternative wellbeing models (e.g., MHC, PERMA) that can be used when evaluating wellbeing in the context of pro-environmental actions. These models, some of which are already incorporated into environmental literature while others are overlooked, are designed to be inclusive and considerate of various perspectives. By adopting a positive psychology perspective on wellbeing, this chapter aims to provide a comprehensive understanding of the impact of biodiversity (illustrated through the Let It Bee project) and pro-environmental behaviours on wellbeing, ensuring that all voices are heard and valued.

As we delve into various wellbeing theories, ranging from the Mental Health Continuum (MHC) to concepts like Flourishing and the PERMA model, it becomes apparent that mental health encompasses more than just the absence of illness. These theories highlight the complexity of our mental wellbeing, emphasising the importance of positive experiences, meaningful activities, and positive relationships.

Evidence suggests that individuals can attain high levels of both emotional (hedonic) and psychological and social (eudaimonic) wellbeing by fostering a connection with nature (Capaldi et al., 2014; Pritchard et al., 2020). Those who feel connected to nature are thus more likely to experience flourishing and psychological functioning (Pritchard et al., 2020). This is significant because flourishing individuals are more inclined to demonstrate empathy and care towards others (Weinstein et al., 2009) and the environment (Kurth et al., 2020). This emerging area of interest will be revisited to explore how our relationship with the environment influences our behaviours towards it and, reciprocally, how these behaviours impact our mental wellbeing.

Moreover, research indicates that our subjective wellbeing (SWB) is enhanced when we connect with nature. More than merely spending time in nature is required; we should engage with nature, appreciating all it offers (Richardson et al., 2021). Simple acts like smelling flowers can elevate our SWB (Richardson et al., 2021). These acts may also induce a flow state wherein we are wholly immersed in and perceive ourselves as integral parts of nature.

While connectedness to nature and seeing ourselves as part of it are pivotal for our mental wellbeing, we will now examine other environmental factors that influence our mental health.

Introduction to illbeing

The concept of wellbeing is a multifaceted and often perplexing term in academic literature, even down to its spelling variations such as "wellbeing," "well-being," or "well being." Compounding the complexity, wellbeing is perceived and approached from diverse disciplinary perspectives, including anthropology, sociology, and psychology.

Broadly speaking, from the psychological viewpoint, wellbeing can be understood through either a deficit or an abundance perspective. In the deficit perspective, wellbeing is viewed as the absence of pathological states; individuals are considered to have wellbeing if they are not diagnosed with mental health disorders such as anxiety, depression, borderline personality disorder, and schizophrenia, among others. However, the emergence of positive psychology has shifted the focus towards an abundance perspective of wellbeing, emphasising the presence of positive outcomes and that make life worth living and contribute to the state of psychological flourishing.

In this chapter, we will initially examine wellbeing from the standpoint of illbeing, which is defined by the absence of mental disorders. Specifically, we will explore eco-anxiety, which is the main concept discussed in relation to pro-environmental behaviour. Subsequently, we will delve into the abundance perspective of wellbeing, examining the various elements and outcomes that contribute to individuals leading fulfilling and meaningful lives.

Eco-anxiety and wellbeing

In environmental psychology, there is significant discussion surrounding a phenomenon known as eco-anxiety, which has emerged as a consequential aspect of environmental stewardship. Eco-anxiety captures the anxious responses of individuals facing evolving environmental challenges. Pihkala (2020) suggests that this concept goes beyond just climate anxiety, encompassing a wider range of concerns related to ecological crises. Consequently, those dealing with eco-anxiety find themselves immersed in feelings of apprehension, foreboding, and pessimism due to environmental degradation.

What makes eco-anxiety particularly intriguing is its nuanced connection to diagnosable mental health conditions, such as depression or anxiety (Daouda et al., 2021; Wullenkord et al., 2021). This association, varying from minor to moderate, indicates that while eco-anxiety is related to mental health issues, it also stands as a distinct experience for some individuals. Thus, there are cases where individuals may face eco-anxiety on its own, without accompanying forms of anxiety. This highlights the localised impact and unique nature of eco-anxiety in psychology. As we explore the interplay between environmental concern and mental wellbeing, it is important to remember that eco-anxiety is a significant and intricate aspect of human experience.

In a recent study, Passmore and colleagues (2023) present a compelling argument, suggesting that the consequences of environmental degradation extend well beyond ecological concerns. By examining various components of wellbeing, the authors illustrate how environmental degradation undermines aspects crucial to our psychological wellbeing, including identity, happiness, and life's purpose and meaning.

According to Passmore et al. (2023), the erosion of identity is a crucial consequence of witnessing environmental degradation. Our belief in humanity's inherent goodness becomes challenged when faced with the reality of environmental destruction. Fuelled by eco-guilt, this challenge leads to existential questioning about our existence and humanity's nature, resulting in what the authors describe as an "identity crisis", which manifests itself as eco-anxiety.

In relation to happiness, Passmore et al. (2023) intricately link eco-anxiety with Ryan and Deci's (2017) theory of motivation, suggesting that individuals derive joy and fulfilment from their connection with nature. The loss of biodiversity, climate crisis, and ecological degradation are significant contributors to profound unhappiness, because they get in the way of that human-nature connection. Importantly, this perspective recognises happiness as multifaceted, with environmental loss representing just one dimension of the broader discourse.

Furthermore, the authors shed light on eco-anxiety's impact on the meaning of life. Drawing from Steger's (2012) definition of meaning, they explain that experiences of nature's beauty play a fundamental role in shaping life's meaning. However, when individuals confront the harsh realities of environmental degradation, understanding other people's motives behind such actions becomes challenging. This leads them to pondering over existential questions that overshadow their life's meaning. In essence, eco-anxiety emerges as a signficant force capable of negatively influencing fundamental aspects of wellbeing, including identity, happiness, and the search for meaning.

Introduction to wellbeing

Navigating the expansive field of wellbeing can be daunting due to the plethora of models and frameworks available. A primary framework that helps us understand wellbeing from both philosophical and psychological perspectives is the classification of hedonic and eudaimonic wellbeing (Disabato et al., 2016). This framework, originating from philosophical roots, was later deconstructed by psychologists into measurable components, laying the groundwork for the wellbeing models widely employed in wellbeing research and practice today.

Despite their diverse origins, all current models share a common characteristic – they are componential. Drawing from philosophy, psychology, sociology, anthropology, and other disciplines, these wellbeing models have been developed with components reflecting the authors' beliefs regarding what constitutes wellbeing. For example, the SWB model, aligned with hedonic wellbeing, encompasses three crucial components: (1) higher levels of positive emotions, (2) lower levels of negative emotions, and (3) life satisfaction (Diener, 1984). Conversely, the psychological wellbeing (PWB) model follows the eudaimonic paradigm, encompassing six components: (1) positive relations with others, (2) environmental mastery, (3) autonomy, (4) a sense of life purpose, (5) self-acceptance, and (6) personal growth.

For years, these two models were compared and contrasted, contributing to confusion about wellbeing as researchers debated which model to adopt. The emergence of positive psychology in the late 1990s aimed to bridge these seemingly opposing paradigms under the umbrella of wellbeing models. These integrated models, often referred to as 'flourishing models' due to their comprehensive nature, represent wellbeing that encompasses both hedonic and eudaimonic perspectives. The components of these models draw from extensive research findings in psychology and other fields, reflecting the evolution of our understanding of wellbeing over the years.

In the context of pro-environmental actions, these wellbeing models can be extended to embrace the connection between individuals and the environment. For instance, positive relations with others may extend to fostering positive relationships with nature, and environmental mastery could involve adopting sustainable practices that contribute to both personal and planetary wellbeing. The evolving understanding of wellbeing aligns with the growing awareness of the interconnectedness between individual wellbeing and environmental health, making these models highly relevant in the context of pro-environmental action and biodiversity.

The models introduced in the context of positive psychology provide structures for understanding and evaluating distinct facets of positive psychological functioning. For example, the Authentic Happiness model (Seligman, 2003), is a model upon which the Positive Ecology model is based. According to Seligman's model, authentic happiness is about living a life of pleasure, engagement, and meaning. The model was highly popularised in the naughties as it was the first positive psychology model to combine hedonic and eudaimonic perspectives. However, a decade later Seligman admitted that happiness was a too simplistic concept to focus on. Instead, we should delve into factors of wellbeing that lead to flourishing, which is why he developed the subsequent PERMA model.

The PERMA Model (Seligman, 2011) comprises a more comprehensive framework for wellbeing. It consists of Positive Emotions, Engagement, Relationships, Meaning, and Accomplishment. Seligman posited that each element of the model is inherently motivating, and integrating all these elements leads to flourishing, representing the highest level of wellbeing. While this model is best known, other models provide evidence of more comprehensive and rigorous research, such as the Mental Health Continuum Model (MHC) model (2002).

The MHC model was developed by Keyes (2002) and incorporates mental health, emotional wellbeing, and social wellbeing. Corey Keyes proposed that this model expands beyond the conventional view of mental health and illness by integrating subjective and psychological wellbeing. Most of the psychological research on flourishing uses this model to measure it.

Another less-known model is the Flourishing Model developed by Diener et al. (2010). This framework offers a comprehensive wellbeing perspective by including vitality among its components. Huppert and So (2013) also assessed Flourishing in Europe across tens of thousands of participants. This model is considered a mixture of SWB (feeling well) and PWB (functioning well). The inclusion of this model in our discussion highlights the global applicability of wellbeing models and their adaptability to different cultural contexts.

These models, representing only a fraction of positive psychology-informed frameworks, are not just theoretical constructs. They are practical tools designed to explore and enhance various facets of wellbeing. Researchers and practitioners can choose a model based on its specific components and focal points, gaining insights into diverse dimensions of positive psychological functioning. This knowledge empowers them to make informed decisions and design effective interventions. Let us explore some of the leading models models in the context of pro-environmental action and biodiversity, demonstrating their practical application.

The Wellbeing Model (PERMA)

Let us now delve into one of the most prevalent models of flourishing in positive psychology, the PERMA model (Seligman, 2011), which identifies five essential elements crucial for experiencing flourishing and living a fulfilling life. We will explore it in the context of environmental actions. Seligman posits that individuals are inherently driven to pursue these elements. When all five are present, they operate at the highest possible psychological and emotional level.

Positive Emotion (P): This component encompasses a range of emotions such as joy, gratitude, serenity, interest, hope, pride, amusement, inspiration, and awe. In the context of pro-environmental action, Positive Emotion evaluates the frequency and intensity of positive emotions individuals experience when actively contributing to environmental causes. For instance, participating in tree-planting initiatives or witnessing the positive impact of conservation efforts on biodiversity can evoke these positive emotions.

Engagement (E): Encouraging individuals to fully immerse themselves in activities aligned with their strengths and values, Engagement may evoke the state of "flow" experienced during climate action. This includes actively participating in sustainability projects and demonstrating a profound connection to tasks such as advocating for renewable energy solutions.

Relationships (R): Recognising the significance of positive and meaningful connections, the Relationships component evaluates the quality of social interactions formed through collaborative pro-environmental efforts. This might involve assessing the impact of community engagement in climate change mitigation projects or the camaraderie developed during environmental awareness campaigns.

Meaning (M): Pursuing Purpose, Meaning, is significant for wellbeing. In the context of biodiversity, Meaning explores whether individuals perceive their actions, such as habitat preservation or wildlife conservation, as purposeful. This may contribute to a clear sense of direction and meaning in their lives.

Accomplishment (A): This component gauges an individual's perception of achieving meaningful goals, experiencing competence, and acknowledging personal accomplishments. It reflects the sense of achievement derived from successfully implementing pro-environmental actions, such as reducing carbon footprints or participating in initiatives to reduce plastic waste.

The PERMA model offers a positive and comprehensive framework for understanding wellbeing, emphasising the presence of positive factors. In the context of environmental and climate action, it serves as a guiding principle for individuals and professionals, facilitating a flourishing and satisfying life centred around meaningful contributions to a sustainable and thriving planet.

Mental Health Continuum

Introduced by Corey Keyes in 2002, the MHC model combines mental illness and mental health. This model encompasses the entire spectrum of mental wellbeing, acknowledging that it extends beyond the mere absence of mental illness. Instead, the continuum spans from languishing to flourishing, capturing the diverse nuances of mental health. Let's explore the three core components that form the MHC model:

Psychological Wellbeing: Revolves around exploring factors that make our lives worth living. They include such concepts as self-acceptance or autonomy. Individuals participating in pro-environmental actions who function at the highest psychological wellbeing level may derive satisfaction from initiating changes that will improve their environment, contributing to an overall state of psychological wellbeing.

Emotional Wellbeing: Focuses on experiencing an emotional balance whereby individuals report more positive than negative emotions daily. Active involvement in biodiversity preservation may lead to positive emotions such as joy, contentment, and a sense of accomplishment.

Social Wellbeing: Refers to the quality of social connection that extends from one-to-one relationships though to communities and societies. Engaging in pro-environmental actions may foster positive social connections and relationships within the community. Collaborative efforts for environmental conservation may contribute to a sense of belonging and active social engagement, aligning with the social wellbeing dimension of the MHC model.

Flourishing model

As conceptualised by Diener and colleagues in the model created in 2010, Flourishing represents a holistic approach to wellbeing that spans various dimensions. This model encapsulates several key components, each playing a pivotal role in contributing to an individual's sense of flourishing:

Competence reflects an individual's perceived capability and proficiency in navigating life's challenges and tasks. Demonstrating competence in biodiversity and pro-environmental actions may involve acquiring knowledge about local ecosystems, conservation practices, and sustainable behaviours.

Self-acceptance involves embracing and acknowledging oneself, including strengths and weaknesses, without undue self-criticism. In environmental stewardship, self-acceptance could manifest as acknowledging personal contributions to biodiversity preservation, recognising efforts, and understanding that progress in pro-environmental actions is an ongoing journey.

Meaning means finding significance and purpose in life, transcending individual desires for a broader, meaningful existence. Engaging in actions that contribute to biodiversity conservation and environmental sustainability can fill life with profound meaning as individuals connect with a larger purpose beyond themselves.

Relatedness focuses on fostering positive connections with others, building a sense of community, and maintaining healthy relationships. In pro-environmental activities, relatedness may involve collaborative efforts within a community to address environmental challenges, emphasising the interconnectedness of individuals committed to biodiversity conservation.

Optimism entails maintaining a positive outlook and anticipating favourable outcomes. Optimism in environmental actions involves believing in the potential for positive change, such as trusting that collective efforts can mitigate climate change impacts and preserve biodiversity.

Giving encompasses acts of generosity, altruism, and contributing to the wellbeing of others. Giving in the environmental context may involve supporting initiatives that promote biodiversity, such as participating in tree-planting drives, donating to conservation organisations, or volunteering for environmental clean-up projects.

Engagement involves fully immersing oneself in and investing in activities that align with one's interests and values. Engagement in pro-environmental actions could mean actively participating in ecological restoration projects, advocating for sustainable practices, or leading community initiatives for environmental awareness.

With its multifaceted dimensions, this flourishing model provides a comprehensive framework that could extend beyond individual wellbeing. In the context of biodiversity and pro-environmental actions, these components offer valuable insights into how individuals can lead fulfilling lives while actively contributing to preserving our natural world.

Flourishing in Europe

Public health, as envisioned by Huppert and So (2013) is intricately linked to an individual's overall wellbeing. Their model encompasses core and additional elements that comprise wellbeing.

1 **Core Elements**

Positive Emotion involves experiencing joy, gratitude, contentment, and other uplifting feelings. Positive emotion in biodiversity and environmental activities can manifest through the joy of witnessing ecological restoration, gratitude for a thriving ecosystem, and contentment from contributing to nature conservation.

Engagement involves fully immersed and invested in activities aligned with one's interests and values.*:* Engagement in biodiversity and environmental initiatives includes actively participating in conservation projects, ecological restoration, or sustainable practices, aligning actions with environmental values.

Meaning encompasses finding significance and purpose in life. *Pro-environmental Context:* Meaning in biodiversity and environmental actions is derived from contributing to a larger purpose, such as preserving ecosystems, fostering environmental awareness, and actively participating in conservation efforts.

Apart from the core elements, there are additional elements, three of which individuals need to develop to flourish. They are self-esteem, optimism, resilience, vitality, self-determination and positive relationships. Which three of the additional elements are developed are up to an individual and the values they hold in life.

In summary, the Public Health Model by Huppert and So, enriched by biodiversity and pro-environmental action examples, highlights the interconnectedness between individual wellbeing and the planet's health. Individuals can lead fulfilling lives by embracing positive appraisal, characteristics, and functioning while actively contributing to environmental sustainability and biodiversity conservation.

Psychological richness

The emergence of the concept of psychological richness, proposed by Oishi and Westgate (2022), addresses a gap in existing frameworks for explaining wellbeing and a good life. Moving beyond the traditional dichotomy of eudaimonic (meaning-based) and hedonic (pleasure-based) approaches, the authors introduce a compelling third dimension: psychological richness. This dimension encapsulates a life enriched by diverse, intriguing, and perspective-altering experiences, alongside happiness and meaning, as integral components of a fulfilled existence.

Two recent studies have identified the link between psychological richness and pro-environmental behaviour, highlighting its relevance in promoting sustainable engagement. In a quantitative investigation conducted by Wei et al. (2023), it was found that individuals who led psychologically rich lives showed a greater inclination to engage in environmentally sustainable activities. This highlights the potential significance of the diversity and depth of personal experiences in fostering proactive environmental stewardship.

Additionally, in a qualitative study by Burke and Corrigan (2024), researchers emphasised that participation in pro-environmental behaviours resulted to participants' experiences of psychological richness. They further suggested that involvement in such actions could serve as an intervention to enhance this dimension of wellbeing. Thus, psychological richness offers an interesting framework for exploring the impact of environmental actions on wellbeing.

Applications of wellbeing in the context of pro-environmental behaviour

Beyond its evident environmental impact, pro-environmental behaviour is now emerging as a catalyst for a cascade of psychological, emotional, and social wellbeing benefits. A recent surge in research is uncovering the nuanced connections between eco-friendly actions and the broader spectrum of psychological flourishing.

The Bee Well project

The Bee Well project was initiated to investigate the impact of pro-environmental behaviours, particularly beekeeping, on the wellbeing of farmers, their families, and the wider community. Distinguished as the pioneering pro-environmental initiative to delve into wellbeing using various positive-psychology models, the project yielded insightful findings.

Our research (Burke & Corrigan, 2024) revealed that engaging in beekeeping, aimed at enhancing biodiversity and water quality in the area, led to a multitude of positive outcomes across different dimensions of wellbeing, as outlined in prominent models such as PERMA (Seligman, 2011), the MHC (Keyes, 2002), and Psychological Richness (Oishi & Westgate, 2022).

Of particular significance was the observation that participants, inspired by the bee societies, shifted their behaviour and outlook towards others. This psychological richness, typically associated with experiences like travel, club memberships, or sports (Oishi et al., 2021), was extended in this study to beekeeping, highlighting its potential as a source of profound psychological enrichment.

While our findings offer promising insights, further research is needed to compare the impact of beekeeping on psychological richness with other activities, especially within the context of environmental sciences. Given the established correlation between beekeeping and pro-environmental behaviours (Wei et al., 2023), exploring this aspect can deepen our understanding of how nature-based activities contribute to overall wellbeing.

Moreover, the implications of our findings extend beyond theory and towards practical realms of the future of pro-environmental actions. Over the years, many Positive Psychology Interventions have been developed to bolster elements of wellbeing (e.g., Pawelski, 2020; Burke et al., 2023). Our study suggested that beekeeping has the potential to emerge as an important Positive Psychology Intervention, offering individuals and their families an avenue to improve their psychological, emotional, and social wellbeing while simultaneously fostering environmental stewardship.

The Bee Well project shed light on the symbiotic relationship between pro-environmental behaviours and wellbeing. It highlighted the transformative potential of nature-based interventions in nurturing flourishing individuals and communities.

Wellbeing literacy vis-à-vis human relationship with nature

A novel concept known as "wellbeing literacy" has been introduced to show the impact of the language individuals use that can either enhance or reduce their overall wellbeing. Pioneered by Oades and her colleagues (2021), this emerging idea is deemed a crucial missing link in wellbeing education.

At its core, wellbeing literacy is envisioned as the proficiency to understand and craft language when describing one's wellbeing journey. It transcends mere vocabulary, extending to a nuanced comprehension and an ability to use various wellbeing models. This entails articulating thoughts, feelings, and experiences related to wellbeing verbally and in writing. Moreover, wellbeing literacy involves cultivating an awareness of how the environmental context influences our overall wellbeing, fostering the creation of habits around discussing and expressing oneself regarding wellbeing. In essence, wellbeing literacy encapsulates the multidimentional ability to navigate the language of wellbeing, offering a comprehensive approach to enhancing our mental and emotional health.

Environmental sciences has become a newfound frontier in a fascinating expansion of the wellbeing literacy concept. Recent work by Lomas (2019) delves into the cross-cultural lexicon of eco-connection within the context of overall wellbeing. Lomas identified three pivotal dimensions of eco-connection.

Firstly, a prevalent theme emerges, emphasising the sacrality of nature. This extends beyond mere appreciation of nature, touching upon fundamental human needs such as meaning, breathing, and life. Exploring this dimension connects nature and various mythologies, including Greek and Roman narratives, with a nod to nature's inherent ability to create and nurture life.

The second thematic thread delves into the profound bonds individuals form with nature. Specifically, how individuals articulate their intertwining with the natural world, describing a sense of rootedness and experiencing a deep longing. These personal stories highlight the deep connection between humanity and the environment.

Finally, many words revolve around appreciation of nature. This encompasses activities like savouring nature through practices like forest bathing, engaging all senses to sense nature's presence, and relishing the aesthetics and beauty inherent in the natural world. In the context of wellbeing literacy, these insights become a valuable resource for educating individuals on optimising their connection with nature. Individuals can unlock the profound benefits of connecting with nature by delving into the dimensions of sacrality, bonding, and appreciation. When applied to our relationship with the environment, wellbeing literacy offers a pathway to embrace the eloquence of nature's language, fostering a heightened sense of overall wellbeing.

Positive health

"Positive health" represents a paradigm shift in how we perceive and pursue overall wellbeing (O'Boyle et al., 2024). It transcends the traditional notion of health as merely the absence of disease. Positive health encompasses both a state to be achieved and an ongoing process of growth and fulfilment.

At its core, positive health entails reaching the highest levels of physical, mental, social, emotional, and existential flourishing. It goes beyond the absence of illness to encompass a holistic state of thriving in all aspects of life. Rather than focusing solely on treating ailments, positive health emphasises optimising every dimension of wellbeing to attain a state of optimal living.

Moreover, positive health recognises the continuous and dynamic journey towards wellbeing. It acknowledges that progress towards thriving is an ongoing process characterised by continuous growth, adaptation, and self-discovery. Each step towards enhancing physical vitality, mental resilience, social connections, emotional intelligence, and finding meaning contributes to this positive health journey.

In the context of health, embracing the concept of positive health is crucial for promoting overall wellbeing and preventing illness. By shifting our focus from solely addressing disease to fostering thriving, individuals can proactively engage in behaviours and practices that promote optimal health across all dimensions of life. This approach not only enhances individual wellbeing but also contributes to building healthier communities and societies at large.

Burke and colleagues (2023) define Positive Health as integrating positive psychology and lifestyle medicine research and practices. This concept encompasses investigations that explore the impact of combining lifestyle medicine and positive psychology concepts. Moreover, Burke et al. (2024a) describe positive health interventions as tools designed to cultivate psychological, emotional, intellectual, physiological, and social resources. These interventions not only demonstrate efficacy in improving health and wellbeing but also aim to alleviate the burden of disease. They delve into the interconnectedness of body and mind. For instance, they include activities promoting overall wellbeing that alleviate disease symptoms and enhance physiological health and improve psychological, emotional, and social wellbeing.

In a recent study by Burke and colleagues (2024b), efforts have been made to integrate positive health interventions within the framework of pro-environmental activities. The researchers conducted a study involving post-primary students who participated in a four-week intervention to promote environmental engagement alongside positive psychology interventions. The objective was to evaluate the impact of these interventions on the participants' overall wellbeing. The research results indicated an increase in wellbeing among both groups involved in the intervention, suggesting the effectiveness of combining environmental activities with positive psychology interventions in enhancing wellbeing. This study highlighted the potential synergistic benefits of addressing environmental and psychological dimensions in interventions promoting holistic health and wellbeing.

In summary, as ongoing research unfolds, a positive correlation emerges between pro-environmental engagement and improved wellbeing. While the field is still evolving, existing evidence points to the potential benefits of actively participating in environmental initiatives for one's overall sense of wellbeing. These findings highlight the significance of incorporating environmental stewardship into personal and community practices, not only for the planet's health but also for enhancing individual wellbeing. Further exploration and comprehensive studies are necessary to fully understand the nuanced relationship between pro-environmental behaviours and wellbeing across diverse populations and contexts.

Conclusion

In this chapter, we have examined the concept of individual wellbeing. Initially, our focus was on understanding wellbeing through the lens of ill-being, particularly in the context of eco-anxiety and its implications for pro-environmental behaviour. Subsequently, we delved into the principal models rooted in positive psychology that inform theories of wellbeing, specifically concerning pro-environmental behaviour. Our exploration extended to the wellbeing literacy model and its potential applications, as well as a novel model of Positive Health and its associated research in the realm of pro-environmental behaviour. While there is a scarcity of research directly addressing the models of flourishing in relation to pro-environmental behaviour, existing studies provide insights into the various components of wellbeing, which will be further discussed in subsequent chapters.

References

Britton, E., & Foley, R. (2020). Sensing Water: Uncovering Health and Well-Being in the Sea and Surf. *Journal of Sport and Social Issues*. https://doi.org/10.1177/0193723520928597

Burke, J., & Corrigan, S. (2024). Bee Well: A positive psychological impact of a pro-environmental intervention on beekeepers' and their families' wellbeing. *Frontiers in Psychology (Environmental Psychology)*, 15. https://doi.org/10.3389/fpsyg.2024.1354408

Burke, J., Dunne, P. J., Meehan, T., O'Boyle, C. A., & van Nieuwerburgh, C. (2023). *Positive health: 100+ research-based positive psychology and lifestyle medicine tools to enhance your wellbeing*. Routledge. https://doi.org/10.4324/9781003279959

Capaldi, C. A., Dopko, R. L., & Zelenski, J. M. (2014). The relationship between nature connectedness and happiness: A meta-analysis. *Frontiers in Psychology*, 5, 976.

Daouda, C. M., Blanchard, M. A., Coussement, C., & Heeren, A. (2021). On the measurement of climate change anxiety: French adaption and validation of the Climate Anxiety Scale. *Psychologica Belgica*, 62(1), 123.

Diener, E. (1984). Subjective well-being. *Psychological Bulletin*, 95, 542–575.

Diener, E., et al. (2010). New well-being measures: Short scales to assess flourishing and positive and negative feelings. *Social Indicators Resesearch*, 97, 143–156. https://doi.org/10.1007/s11205-009-9493-y

Disabato, D. J., Goodman, F. R., Kashdan, T. B., Short, J. L., & Jarden, A. (2016). Different types of well-being? A cross-cultural examination of hedonic and eudaimonic well-being. *Psychological Assessment*, 28, 471–482.

Huppert, F. A., & SO, T. T. C. (2013). Flourishing across Europe: Application of a new conceptual framework for defining well-being. *Social Indicators Research*, 110, 837–861.

Hyland, P., et al. (2022). State of Ireland's mental health: Findings from a nationally representative survey. *Epidemiology and Psychiatric Sciences*, 31, e47. https://doi.org/10.1017/S2045796022000312

Keyes, C. L. M. (2002). The mental health continuum: From languishing to flourishing in life. *Journal of Health and Social Behavior*, 43, 207–222.

Kurth, A. M., Narvaez, D., Kohn, R., & Bae, A. (2020). Indigenous nature connection: A 3-week intervention increased ecological attachment. *Ecopsychology*, 12(2), 101–117. https://doi.org/10.1089/eco.2019.0038

Lomas T. (2019). The elements of eco-connection: A cross-cultural lexical enquiry. *International Journal of Environmental Research and Public Health*, 16(24), 5120. https://doi.org/10.3390/ijerph16245120

Oades, L. G., Jarden, A., Hou, H., Ozturk, C., Williams, P., R Slemp, G., & Huang, L. (2021). Wellbeing literacy: A capability model for wellbeing science and practice. *International Journal of Environmental Research and Public Health*, 18(2), 719. https://doi.org/10.3390/ijerph18020719

O'Boyle, C. A., Lianov, L., Burke, J., Frates, B., & Boniwell, I. (2024). Positive health: An emerging new construct. In J. Burke, I. Boniwell, B. Frates, L. Lianov, & C. A. O'Boyle (Eds.), *The Routledge international handbook of positive health* (pp. 2–23). Routledge.

Oishi, S., Choi, H., Liu, A., & Kurtz, J. (2021). Experiences associated with psychological richness. *European Journal of Personality*, 35(5), 754–770. https://doi.org/10.1177/0890207020962

Oishi, S., & Westgate, E. C. (2022). A psychologically rich life: Beyond happiness and meaning. *Psychological Review*, 129(4), 790–811. https://doi.org/10.1037/rev0000317

Passmore, H.-A., Lutz, P. K., & Howell, A. J. (2023). Eco-anxiety: A cascade of fundamental existential anxieties. *Journal of Constructivist Psychology*, 36(2), 138–153. https://doi.org/10.1080/10720537.2022.2068706 Pawelski, J. O. (2020). The elements model: Toward a new generation of positive psychology interventions. *The Journal of Positive Psychology*, 15(5), 675–679. https://doi.org/10.1080/17439760.2020.1789710

Pihkala, P. (2020). Anxiety and the ecological crisis: An analysis of eco-anxiety and climate anxiety. *Sustainability*, 12, 7836.

Pritchard, A., Richardson, M., Sheffield, D., & McEwan, K. (2020). The relationship between nature connectedness and eudaimonic well-being: A meta-analysis. *Journal of Happiness Studies*, 21(3), 1145–1167. https://doi.org/10.1007/s10902-019-00118-6

Richardson, K. et al. (2023). Earth beyond six of nine planetary boundaries. *Science Advances*, 9(37), eadh2458. https://doi.org/10.1126/sciadv.adh2458

Richardson, M., Passmore, H.-A., Lumber, R., Thomas, R., & Hunt, A. (2021). Moments, not minutes: The nature-wellbeing relationship. *International Journal of Wellbeing*, 11(1). https://doi.org/10.5502/ijw.v11i1.1267

Ryan, R. M., & Deci, E. L. (2017). *Self-determination theory: Basic psychological needs in motivation, development, and wellness*. Guilford Press.

Ryff, C. D. (1989). Happiness is everything, or is it? explorations on the meaning of psychological well-being. *Journal of Personality & Social Psychology*, 57, 1069–1081.

Seligman, M.E.P. (2003). *Authentic happiness. Using the new positive psychology to realize your potential for lasting fulfillment*. Free Press.

Seligman, M. (2011). *Flourish: A new understanding of happiness and well-being – and how to achieve them*. Free Press.

Steger, M. F. (2012). Experiencing meaning in life: Optimal functioning at the nexus of well-being, psychopathology, and spirituality. In P. T. P. Wong (Ed.), *The human quest for meaning: Theories, research, and applications* (2nd ed., pp. 165–184). Routledge/Taylor & Francis Group.

Tost, H., Champagne, F. A., & Meyer-Lindenberg, A. (2015). Environmental influence in the brain, human welfare and mental health. *Nature Neuroscience*, 18(10), 1421–1431. https://doi.org/10.1038/nn.4108

Wei, X., Yu, F., Peng, K., & Zhong, N. (2023). Psychological richness increases behavioral intention to protect the environment. *Acta Psychologica Sinika*, 55(8), 1330–1343. https://doi.org/10.3724/SP.J.1041.2023.01330

Weinstein, N., Przybylski, A. K., & Ryan, R. M. (2009). Can nature make us more caring? Effects of immersion in nature on intrinsic aspirations and generosity. *Personality and Social Psychology Bulletin*, 35(10), 1315–1329. https://doi.org/10.1177/0146167209341649

Wullenkord, M. C., Tröger, J., Hamann, K. R. S., Loy, L. S., & Reese, G. (2021). Anxiety and climate change: A validation of the Climate Anxiety Scale in a German-speaking quota sample and investigation of psychological correlates. *PsyArXiv*. https://psyarxiv.com/76ez2/

5 Individual emotional wellbeing

In his mid-40s, John found solace in simply immersing himself in nature and biodiversity without needing to identify every species. He had endured a challenging period in his life, facing a marriage separation and financial strain. Walking alongside a river enveloped by trees and nature became his refuge, a way to unwind and find balance amidst the turmoil.

John's journey to nature was not a deliberate choice but a fortunate one. Initially, he set out on morning walks to improve his physical fitness, accompanied by the constant chatter of news channels through his headphones. However, he soon realised that this was not the solace he sought. It was a day when his phone battery died mid-walk that he discovered the therapeutic power of nature. Without the distraction of news, he found himself attuned to the natural world around him, and for the first time in a long while, he felt a genuine sense of enjoyment.

John's decision to intentionally disconnect from news during his walks marked a significant turning point in his relationship with nature. Gradually, he transitioned from a passive observer to an active appreciator of the natural world, free from external distractions. This daily hour of unplugged walking became a vital ritual for him to reset his mindset and prepare for the day ahead. He found himself not just physically refreshed but also mentally rejuvenated, less stressed, and better equipped to face the challenges of work.

John's decision to immerse himself in a biodiverse environment during his walks profoundly impacted his overall wellbeing. It helped him find a balance between work and home life and heightened his awareness of his surroundings in other environments. His daily encounters with nature deepened his appreciation for biodiversity, a perspective that he carried with him beyond his walks.

John's experience is a powerful testament to the therapeutic power of nature. It highlights the importance of carving out time to disconnect from technology and reconnect with the natural world, especially during stress and upheaval. His story offers reassurance that nature can provide comfort and healing.

DOI: 10.4324/9781003452676-7

In the quest for personal wellbeing, individuals are driven by a primal instinct deeply ingrained within their evolutionary heritage, alongside inner motivations that shape their aspirations and actions. The connection between physical health and overall wellbeing has compelled individuals to actively seek happiness and strive for increased hedonic pleasure and eudaimonic fulfilment. This pursuit, however, has its complexities in the context of pro-environmental behaviour and challenges, as we shall explore in this chapter, making our discussion all the more relevant and significant.

Despite the relentless quest for happiness, studies by Mauss et al. (2011, 2012) suggest that pursuing happiness can sometimes lead to unintended consequences, such as feelings of isolation and a decline in overall happiness. Consequently, emotional wellbeing remains inherently precarious, subject to the twists and turns of life's journey. These findings, though paradoxical, offer valuable insight into the complexities of emotional wellbeing, potentially reshaping our understanding of personal fulfilment.

The current chapter explores the intricate dynamics of emotional wellbeing, particularly within the context of pro-environmental behaviour. By exploring how environmental pursuits intersect with individual wellbeing, we seek to clarify how embracing environmental causes can enhance emotional wellbeing. By fostering a deeper connection to nature and promoting a sense of purpose beyond self-interest, environmental engagement can not only enrich the quality of individuals' lives but also offer a meaningful alternative to the pursuit of happiness.

Emotional wellbeing

The foundation of emotional wellbeing theory discussed in this chapter draws from the positive psychological perspective. This perspective aligns closely with the theory of Subjective Wellbeing proposed by Diener (1984), which posits that individuals achieve a state of wellbeing when three essential conditions are met. Firstly, they report more robust experiences of positive emotions; secondly, they experience reduced levels of negative emotions; and thirdly, they report satisfaction with their lives.

Within this framework, the first two components are centred on emotions, which are not related to each other. Therefore, any efforts to cultivate emotional balance must consider positive and negative emotions in tandem. However, this chapter will primarily focus on exploring positive emotions. This choice stems from the fact that significantly less is known about positive emotions compared to negative ones.

Focusing on positive emotions in this chapter allows us to explore the lesser-known facets of emotional experience. Despite their understudied nature, positive emotions are crucial in shaping individuals' experiences and perceptions of happiness, fulfilment, and contentment. By understanding the nuances of positive emotions, we can gain valuable insights into strategies for enhancing emotional wellbeing in the context of pro-environmental behaviour and fostering a more fulfilling life.

Positive emotions

In recent decades, psychology has primarily focused on understanding negative emotions and their evolutionary purpose in supporting our survival. Negative emotions have been extensively researched, given their role in helping individuals address immediate problems. Positive emotions, on the other hand, have often been overlooked and dismissed as simply contributing to momentary happiness and joy.

Consequently, extensive research has delved into the intricacies of negative emotions, aiming to explore their meanings, origins, and significance. While negative emotions have been highly regarded for their role in our survival, positive emotions have often been overlooked and dismissed. Even prominent models of emotions, such as Paul Ekman's model based on facial expressions, gravitated towards understanding the nuances of negative emotions. Both types of emotions are significant; understanding and embracing negative emotions can foster our personal growth and make us more effective in reacting to the outside world. However, this chapter will centre its focus on positive emotions, exploring their role, function, and impact on health and wellbeing.

In the 1970s, a paradigm-shifting moment shattered the prevailing narrative about positive emotions. Dr Ison's ground-breaking series of experiments demonstrated their utility across various facets of life. These studies recognised positive emotions as serving more significant purpose in the decision-making process, challenging the notion that their sole purpose was pursuing happiness.

Broaden-and-Build theory

Drawing upon an expanding body of research, Professor Barbara Frederickson formulated the Broaden-and-Build theory of positive emotions (Fredrickson, 2001). This theory highlights the diverse advantages of positive emotions. Primarily, positive emotions are seen as expanding our perspective, literally and metaphorically, enhancing cognitive abilities. A study by Rowe and colleagues (2007) employing sophisticated eye-tracking technology revealed that inducing positive emotions expanded participants' peripheral vision. This broadening effect heightened cognitive flexibility and helped participants unlock their creative potential and improve problem-solving skills, which is vital for effectively addressing environmental challenges.

Apart from problem-solving, Kuhbandner and colleagues (2011) showed the profound impact of positive affect on encoding incoming information into iconic memory among participants. Encoding environmental information into iconic memory could shape attitudes, emotions, and behaviours, influencing attention, awareness, and emotional responses. Vivid images can raise awareness of environmental issues, evoke emotions like empathy, and encourage pro-environmental actions. Iconic memory also enhances retention and can foster a sense of connection with nature, leading to increased engagement in conservation efforts and sustainable behaviours.

Picture experiencing stress and fear. Your body responds by tensing up, diverting resources from systems like digestion and reproduction to fuel your brain for problem-solving and pumping adrenaline to prepare for fight or flight. This heightened state narrows your vision, limiting peripheral sight. Furthermore, to aid cognitive function, your mind presents only a few options for addressing the issue. Offering too many choices could induce indecision, hindering rather than aiding survival.

On the other hand, when experiencing positive emotions, your body becomes receptive to the subtleties of the environment, facilitating an expansion of peripheral vision. This heightened awareness broadens your physical perception and opens the mind to a spectrum of creative ideas, enhancing problem-solving abilities in the process. Positive emotions create an atmosphere conducive to cognitive flexibility, encouraging innovative thinking and fostering a broader range of potential solutions to the challenges. In this state, the mind is freed from constraints, allowing for the exploration of diverse perspectives and approaches, ultimately leading to more effective and creative problem resolution.

Being receptive to the subtleties of the environment and expanding one's vision can enhance awareness of ecological issues and the importance of biodiversity. This heightened awareness fosters a deeper appreciation for nature, motivating individuals to take proactive steps to protect it. Additionally, becoming more creative in problem-solving enables the development of innovative solutions to environmental challenges, facilitating practical pro-environmental actions and conservation efforts. Overall, these qualities could contribute to sustainable practices and promote the preservation of biodiversity for future generations.

Another significant benefit is the ability for positive emotions to build psychological resources like hope and optimism. This is why individuals with the capacity to experience higher levels of positive emotions demonstrate greater resilience in the face of adversity, contrasting with the inhibiting effects of prolonged negative emotions. Frederickson describes this as an "upward spiral", highlighting the enduring positive impact of positive emotions on overall wellbeing.

Contrary to popular belief, positive emotions go beyond temporary happiness and impact people long-term. For instance, individuals who regularly experience positive emotions, such as joy, gratitude, and contentment, tend to have lower levels of stress hormones like cortisol, which can lead to improved cardiovascular health. Additionally, positive emotions have been linked to better immune function, reducing the risk of illness and promoting overall physical wellbeing.

Furthermore, positive emotions can have a profound impact on mental health, fostering psychological resilience and buffering against the negative effects of stress and anxiety. For example, individuals who experience frequent positive emotions are more likely to engage in adaptive coping strategies, such as seeking social support and reframing challenging situations in a more positive light. This resilience can contribute to better mental health outcomes, including reduced symptoms of depression and anxiety. Pro-environmental actions can bolster long-term wellbeing by nurturing these essential resources when framed positively.

Additionally, positive emotions exhibit an "undoing" effect. At the same time, negative emotions keep the body in a heightened state of alertness, which can potentially result in adverse consequences. Positive emotions alleviate stress, restore equilibrium, and contribute to long-term wellbeing. A pivotal study conducted with university students highlights the lasting influence of positive emotions. Participants who participated in activities that fostered positive emotions exhibited notably reduced levels of worry the next day after being subjected to an experiment inducing negative emotions, compared to those who dealt with negative emotions without balancing them up with positive emotions (Bahrami et al., 2012). This study emphasised the significance of proactively addressing negative emotions and neutralising them with positive emotions.

For instance, consider a scenario where a group of individuals gathers at an environmental rally. Throughout the presentation, speakers delve into the grim reality of the planet's degradation, emphasising the irreversible damage caused by human actions. If this is the message conveyed, attendees may go home feeling overwhelmed with worry. This sentiment lingers long after the event. However, concluding the rally with a burst of positive emotions, such as sharing words of hope for change or celebrating the collective effort of attendees to enact positive transformation, has the potential to counteract the negative impact of despair. These positive emotions mitigate worry, which is often passive, while inspiring individuals to take action towards environmental preservation.

Positive emotions and the environment

In the realm of pro-environmental actions, acknowledging the importance of positive emotions is paramount. While discussing challenges and potential adverse outcomes, leaders should exercise a duty of care, ensuring followers experience positive emotions to counterbalance the inevitable negative aspects of pro-environmental activism. This not only contributes to individual wellbeing but also fosters a positive and creative mindset crucial for effective environmental problem-solving.

Expanding our understanding of positive emotions goes beyond simplistic notions of happiness and joy. Emotional complexity allows individuals to experience a spectrum of emotions simultaneously, fostering emotional intelligence. This richness enables individuals to navigate complex situations, experiencing both despair and hope or sadness and happiness concurrently, which is particularly relevant in environmental contexts.

Eco-Loving-Kindness Meditation: nurturing compassion for nature

Start your journey by grounding yourself. Become aware of the space where you sit and accept it. Feel the floor under your feet or the chair supporting your back.

Now, take five deep breaths to help you centre.

Please think of the beauty of nature, a successful pro-environmental action you participated in, or any other positive environmental changes instigated in the world to safeguard our planet. Consider the positive emotions you experienced at the time and how recalling these moments helps you feel today. Focus on the joy, serenity, relaxation, awe, or any other emotions you experience. Let them wash over you. Feel them in all parts of your body, mind, feet, belly, and other places.

Now, think of something that evokes neutral emotions in you. This could be the parking lot in your local grocery store, with recycling bins recently installed, or the work you have to do in your garden to improve biodiversity. Consider any other pro-environmental actions that had to be done to progress eco-awareness. Now, try to send good wishes to those who made it happen. Feel the gratitude and kindness towards them. Let these feelings wash over you as you experience a burst of positive emotions.

Finally, think of something that frustrates you about pro-environmental stewardship, like people who do things that are not good for the environment. Instead of dwelling on the negative emotions, shift your attention to hope for change or compassion. Allow yourself to relax despite your frustration and express self-love and self-kindness washing over you alongside the frustration. These two types of emotions, both intense, can live symbiotically in each one of us. Remember that positive emotions are still there, even when you are frustrated and all you can think of is that frustration.

Now, take five deep breaths and return to your seat. Hear the sounds and feel the floor beneath you.

In the domain of environmental stewardship, the recognition of positive emotions is important. When broaching the hurdles and potential pitfalls inherent in such discussions, leaders are responsible for cultivating an atmosphere filled with positivity, countering the inevitable negativity of daily lives.

This nurtures individual wellbeing and kindles the fires of a constructive and inventive mindset essential for navigating the complexities of environmental problem-solving.

Eco-mood repair through creative expression (adapted from Dalebroux et al., 2008)

When faced with the distressing realities of environmental degradation, whether from climate crises or biodiversity loss, consider a unique path to emotional healing rooted in eco-consciousness. Take a moment to gather drawing supplies and eco-friendly paints, such as oils or acrylics, each stroke reflecting your dedication to environmental care. Immerse yourself in the creative process, letting your artistic passion flow as you sketch or paint.

Use this creative outlet to craft an image that revitalises your spirit, infuses the positivity of pro-environmental enthusiasm – whether it's the joy of vibrant colours, the peace of serene landscapes, the wonder of nature's beauty, the pride of environmental advocacy, or the calm of contentment. Without limitations, allow your eco-creative expression to unleash your imagination and portray the wellbeing of our planet.

Positive emotions, in particular, the self-transcendent ones like awe, compassion, and gratitude, are increasingly recognised for their potential to inspire pro-environmental behaviour (Zelenski & Desrochers, 2021).

Cultivating positivity for pro-environmental resilience

In environmental stewardship and biodiversity conservation, the significance of positive emotions transcends mere upliftment of mood. Drawing from the profound insights of Fredrickson (2001), a positive emotional state emerges as fertile soil for nurturing essential intellectual, physical, social, and psychological resources. These resources, ranging from problem-solving acumen to cardiovascular health and relationship-building prowess to resilience and optimism, form the bedrock for holistic engagement in pro-environmental endeavours.

Resilience

Resilience refers to the ability to adapt effectively to changing circumstances, particularly in the face of adversity (Masten et al., 1990). It involves the speed at which one can adjust, with more resilient individuals typically bouncing back quicker than others (Davidson & Begley, 2012). Resilience can lead to three primary outcomes: recovery, resistance, or reconfiguration (Lepore &

Revenson, 2006). Recovery involves bouncing back after an adverse event, while resistance entails having protective traits that shield against such events. Reconfiguration occurs when a person changes due to trauma, potentially becoming stronger, healthier, or positively altering their thoughts and behaviours.

Emotional style refers to the consistent patterns of emotional responses and regulatory strategies that individuals exhibit across various situations and over time (Davidson & Beagley, 2012). This research emphasises the role of individual differences in emotional functioning. It highlights how these differences influence psychological wellbeing and overall life outcomes.

Davidson's work suggests that emotional style is not fixed but can be shaped by experiences, environmental factors, and intentional practices. He identifies six dimensions of emotional style, and resilience is one of them. It reflects the ability to bounce back from adversity and maintain emotional equilibrium in the face of challenges.

The neurological processes that underpin resilience are a dynamic interplay between the prefrontal cortex, responsible for rational functions, and the amygdala, a primal and reactive part of the brain. Resilience is closely tied to the prefrontal cortex's ability to regulate and calm the amygdala's response to stress or threats. The swifter this regulation occurs, the more resilient and rational individuals tend to be in the face of adversity. This intricate interaction highlights the importance of cognitive control and emotional regulation, equipping individuals to cope with challenging situations.

Recently, the concept of collective or team resilience has gained attention in research, especially in the context of environmental activism. Collective resilience refers to a team's ability to navigate challenges, crises, and changing circumstances together, adapting effectively and bouncing back as a group (Stoverink et al., 2020). This concept is particularly pertinent in environmental activism, where groups often face complex and daunting challenges. Understanding and fostering collective resilience can enhance the effectiveness and sustainability of environmental activist groups as they work towards their goals.

Collective resilience encompasses two key components (Berkes & Ross, 2013). The first involves fostering adaptive relationships and community learning, uniting individuals, and providing solutions to shared challenges. This component relies highly on individuals experiencing positive emotions that facilitate their connection with others. The second centres on cultivating collective agency and self-organisation, leveraging the team's strengths, values, leadership, and expertise.

As we navigate the complexities of climate change, nurturing positive emotions becomes a critical strategy, fostering a cognitive terrain ripe for innovative and effective pro-environmental endeavours. Positive emotions can undo the effect of negativity, open up our minds unlocking avenues of creativity that might lie dormant during anxiety or anger. Finally, they can build psychological resources, such as resilience, that can help us cope with climate change more effectively. This heightened cognitive flexibility empowers individuals to conceive inventive strategies and initiatives to tackle environmental challenges.

Intensely positive experience (inspired by Burton & King, 2004)

Writing about intensely positive experiences can amplify positive emotions and positively impact health. In a ground-breaking experiment by Burton and King (2004), participants wrote about their most intense positive experiences over three consecutive days. Three months later, follow-up assessments revealed increased levels of wellbeing and significantly fewer doctor visits among participants, accompanied by an improved positive mood. Here is how you can complete this activity, in the context of pro-environmental stewardship and biodiversity.

1. **Recall Moments of Positive Emotions:** Reflect on your most profound encounters with nature, biodiversity, or pro-environmental actions. Whether it was a wilderness experience, a significant conservation initiative, or a realisation of interconnectedness between humans and nature, revisit these memories.
2. **Immerse Yourself in the Experience:** Transport yourself back to those moments, reliving the sensations and insights. Envision the sights, sounds, and scents surrounding you, rekindling feelings of wonder and awe.
3. **Express in Detail:** Capture your reflections in words, painting a detailed picture of your experience. Describe the emotions, thoughts, and revelations you experienced during that transformative encounter.

Eco-anxiety and positive emotions

The daunting challenges of climate change and perceived governmental inaction often evoke frustration, triggering worry and anxiety that may escalate into mental health issues (Dodds, 2021). Positive emotions are particularly useful in this context, counteracting the effect of negative emotions and helping individuals face their challenges with resilience.

A disparity exists between some environmental activists' coercive tactics of guilt and fear, often leading to temporary compliance and the sustaining power of hope and optimism that drives lasting change. This perspective highlights the transformative potential of fostering positive emotional states, guiding individuals towards proactive engagement with the natural world (Carter, 2011), and increasing interest in pro-environmental actions when used wisely.

Recent research highlighted the intricate relationship between eco-anxiety and emotional experiences (Lutz et al., 2023). A daily diary study involving undergraduate students showed that on days marked by surges in eco-anxiety, there was a corresponding imbalance characterised by increased negative emotions and decreased positive ones.

Positive and negative emotions are not inherently correlated (Diener, 2010). Therefore, individuals may experience both increased positive emotions and negative emotions simultaneously. In such cases, positive emotions can assist individuals in coping with negative emotions more effectively. However, when emotional experiences become imbalanced, with high negative and lower positive emotions, individuals may feel overwhelmed and more anxious. This imbalance may have occurred in the scenario described, contributing to heightened eco-anxiety among individuals experiencing such emotional turmoil.

The enduring ripple effects of eco-anxiety lead participants to experience more negative emotions on subsequent days. This is why, when feeling anxious, it is crucial to not only address the root causes of anxiety but also to strive to experience more positive emotions to counterbalance the negative ones. This balanced approach can help individuals manage their anxiety more effectively and promote overall emotional wellbeing.

This finding emphasises the importance of nurturing individuals' emotional wellbeing as they navigate pro-environmental actions. It highlights the imperative to provide individuals with the tools and resources necessary to navigate the ebbs and flows of their emotional journey, thereby mitigating the risk of adverse mental health outcomes.

Gratitude

Within positive psychology, gratitude is one of the most effective interventions for enhancing wellbeing by fostering a sense of wonder, thankfulness, and appreciation for life (Emmons & Shelton, 2005). It transcends mere sentimentality, inviting us to acknowledge life's blessings and cultivate a broader disposition of awareness and appreciation (Wood et al., 2010). Two types of gratitude exist: a trait gratitude and a state gratitude. Trait gratitude is an individual's propensity for being grateful, whereas state gratitude is gratitude that individuals experience following an intervention to count their blessings or by observing others. It is a type of gratitude experienced at a more temporal level.

Trait gratitude towards nature has been linked to the intention to engage in pro-environmental behaviours and actual contributions to environmental causes. Individuals who exhibit higher levels of gratitude towards nature are more likely to express an intention to participate in activities that benefit the environment and are more inclined to donate to environmental organisations (Tam, 2022). Gratitude can be a powerful catalyst for fostering a deeper connection with nature, creating a positive domino effect that influences individuals' relationships with the natural world (Chen et al., 2022).

Furthermore, experimental studies investigating the manipulation of gratitude towards nature have shown some behavioural effects, albeit not consistently robust. While interventions inducing gratitude towards nature have demonstrated the potential to influence pro-environmental behaviour, the observed effects have varied across different studies and contexts (Tom, 2022).

Gratitude, which is seen as the quintessential Positive Psychology Intervention (PPI), is a potent remedy for alleviating depressive and anxiety symptoms and enhancing overall wellbeing (Seligman et al., 2005). Empirical research shows many benefits for those who embrace gratitude: increased life satisfaction, improved psychological functioning, and positive social outcomes (Emmons & McCullough, 2003; Watkins, 2004). Grateful individuals flourish with optimism and withstand challenges such as neuroticism, loneliness, and envy (McCullough et al., 2002). Physically, gratitude even enhances sleep quality for insomniacs (Wood et al., 2009; Digdon & Koble, 2011). Furthermore, the intertwining of gratitude and forgiveness offers solace to those navigating crises, illuminating hope even in the darkest times (DeShea, 2003; Fredrickson et al., 2003).

In interpersonal relationships, gratitude enriches relationships (Algoe et al., 2010). Recipients of gratitude are inspired to reciprocate with affection and support, fostering mutual assistance and selflessness (Bartlett & DeSteno, 2006; Watkins et al., 2006). Gratitude also strengthens community ties, spreading love and appreciation (Fredrickson, 2004; Algoe et al., 2008).

Gratitude is recognised as a driver of both private and public pro-environmental behaviour (Sun et al., 2023). Individuals from lower social classes tend to exhibit a greater propensity for engaging in private pro-environmental actions, such as household resource conservation and personal waste reduction (Bain et al., 2016). Conversely, those from higher social classes demonstrate a stronger inclination towards public pro-environmental behaviours, such as advocacy for environmental policies and participation in community clean-up efforts (Dietz et al., 2005).

This disparity in behaviour is influenced by several factors, including disparities in resource access, cultural norms, and perceptions of social responsibility (Dietz et al., 2005; Bain et al., 2016). Individuals from lower social classes often prioritise actions that directly affect their immediate surroundings and daily lives. In comparison, members of the higher social class may feel a heightened sense of duty to contribute to broader societal and environmental objectives.

Moreover, gratitude emerges as a significant motivator for both types of pro-environmental behaviour. Grateful individuals are more likely to value and appreciate the natural world, recognising its significance for future generations (Emmons & McCullough, 2003). This heightened appreciation can translate into actions taken both privately and publicly to safeguard and conserve the environment.

Understanding the intricate interplay between gratitude, social class, and pro-environmental behaviour offers valuable insights for crafting targeted strategies to promote environmental stewardship across diverse populations (Emmons & McCullough, 2003). By leveraging these factors, policymakers, educators, and environmental advocates can develop interventions that encourage sustainable practices and foster positive attitudes towards environmental conservation (Dietz et al., 2005; Bain et al., 2016).

There is a clear need for further research to understand the mechanisms underlying the relationship between trait gratitude towards nature and pro-environmental behavior. This understanding and the potential impact of experimental manipulations can provide valuable insights into fostering environmentally responsible attitudes and behaviours. Future research could also aim to refine experimental interventions to enhance their effectiveness in promoting sustainable actions.

Practical implementation of gratitude practices

Scholarly research reveals the effectiveness of gratitude interventions, whether implemented as a single activity or integrated into daily life over an extended period (Emmons & McCullough, 2003; Watkins, 2004). However, it is essential to maintain a delicate balance, as excessive repetition and mechanical recording of blessings beyond a specific frequency may diminish their depth and efficacy (Lyubomirsky, 2007; Hefferon & Boniwell, 2011).

Cultivating gratitude for the environment: A path to positive action

The Gratitude List activity, inspired by the 2003 study conducted by Emmons & McCullough, offers a momentary reflection and a continuous practice to explore positive connections with the environment. It serves as a powerful tool, not just for identifying personal gratitude but also for those seeking a "silver lining" (Seligman, 2003) amidst concerns about environmental actions.

Gratitude list

Consider your actions, companions, and positive experiences in your recent weeks relating to the environment, or pro-environmental actions. Note down five things you are grateful for, and for each item, articulate the reasons behind your gratitude. Inspired by Seligman's findings, this practice, where explaining gratitude enhances wellbeing, could create a deeper connection with environmental aspects of our lives.

Three good things

While gratitude practice is beneficial, moderation is key to preventing potential adverse effects (Lyubomirsky, 2007). Start this exercise daily for 1–2 weeks to instil a habit, then transition to a once-a-week routine to maintain its meaningful impact (Hefferon & Boniwell, 2011). Spend ten minutes each night before sleep, listing three positive environmental occurrences and the reasons behind their positivity (Seligman, 2003). Keep a physical record to strengthen the connection with these positive events.

Gratitude letter and visit

The most impactful gratitude activity involves writing a gratitude letter and personally delivering it to the person we appreciate. This action enhances personal happiness and demonstrates sustained improvement for up to a month post-visit (Seligman et al., 2005). Consider expressing gratitude to someone who has positively impacted the environment, ensuring that your acknowledgement is conveyed by handing them the letter or sending it.

Grateful thought substitution

To integrate gratitude into environmental consciousness, proactive thought substitution is recommended. When catching oneself in ungrateful thinking about the environment, we can stop and consciously replace negative thoughts with more grateful ones (Miller, 1995). A four-step technique, identified during counselling sessions, guides the process:

1. Identify ungrateful thoughts.
2. Formulate gratitude-supporting thoughts.
3. Substitute negative thoughts with grateful ones.
4. Act on your newfound grateful perspective.

Gratitude, when applied to environmental contexts, not only aligns with optimism but becomes a catalyst for positive action. By fostering gratitude practices, individuals embark on personal transformation, strengthening their bond with nature and promoting pro-environmental behaviours. This nuanced approach intertwines psychological wellbeing with ecological consciousness, offering a pathway to a more sustainable and harmonious relationship with the environment.

Kindness

Kindness is a fundamental virtue encompassing actions intended to benefit others – a foundational element woven into the fabric of human nature (Curry et al., 2018). From ancient times when survival relied on mutual support within social communities, kindness has evolved into an enduring aspect of the human psyche, albeit with varying degrees of expression.

Various approaches to kindness are witnessed in our daily lives, each paying tribute to humanity's altruistic inclination.

Here are some of the types of Altruism:

- Kin Altruism, an expression of benevolence towards family members, manifests in simple gestures like escorting a child to school or caring for a sick relative.
- Mutualism extends kindness to the broader community through participation in communal efforts, such as collective clean-up projects or aiding a neighbour whose home has been damaged.
- Reciprocal Altruism involves acts of kindness with an eye towards future reciprocity, such as helping an elderly person navigate busy streets or returning a lost wallet to its owner.
- Competitive Altruism, driven by competitive motives, is demonstrated through acts of generosity aimed at enhancing social status or financial benefits, like philanthropic donations for recognition or tax benefits.

Despite the diversity of altruistic acts, one heartening truth is that kindness, acting as a universal remedy, surpasses the constraints of time and place. Embracing this principle, the Dalai Lama shared enduring wisdom: "Be kind whenever possible. It is always possible." This ageless aphorism resonates as a call to collective action, encouraging us to enrich our existence with compassion and goodwill.

Altruism and environmental action

Let us embark upon a journey of exploration, delving into tangible vignettes that illustrate various forms of altruism, particularly within the realm of pro-environmental endeavours:

- Kin Altruism, family members expressing altruism towards each other by children following parents' passion for pro-environmental actions or parents improving biodiverstiy in the area to leave a legacy behind for the future generations.

- Mutualism, community organising local clean-up, nurturing communal gardens or pooling resources to encourage sustainable living practices.
- Reciprocal Altruism, in workplaces, colleagues bonding by facilitating carpooling arrangements, participating in communal recycling programs, or the collective investment in renewable energy solutions.
- Competitive Altruism, individuals and organisations contributing through donations to environmental charities, the patronage of green businesses, or investments in cutting-edge eco-technologies that herald a paradigm shift towards sustainability.

Acts of kindness and environmental actions

Embarking upon the path of pro-environmental stewardship through acts of kindness is a potent and accessible avenue for individuals to weave threads of positivity into human wellbeing. Here, within the crucible of environmentally conscious endeavours, lie myriad ways to enact kindness:

Purposeful purchases

- Instead of conventional tokens of affection, consider eco-conscious alternatives like reusable bags, sustainable homeware, or artefacts crafted from repurposed materials.
- Patronise local artisans and sustainable enterprises, nurturing community bonds and ecology.
- Gift experiences that transcend materialism, such as tickets to ecological events potted vegetables or workshops celebrating sustainable living.

Impactful contributions

- Extend support to environmental charities by donating money.
- Give away unwanted possessions to organisations able to repurpose or recycle them, thus diminishing the waste.
- Crowdfund new eco-friendly initiatives and sustainable paradigms.

Social recycling

- Give away pre-owned and pre-loved items so that others can avail of them.
- Engage in community recycling endeavours where instead of buying new items you can avail of pre-owned products.
- Engage in "swap meets", where you swap unwated items with other people.

Acts of green kindness

- Engage in planting pesticide-free trees, thus leaving a legacy of environmental guardianship.

- Galvanise communal spirit through organised clean-up
- Share the wisdom of eco-conscious living within your social network, encouraging sustainable practices.

Concentrated acts of benevolence

- Designate a day of heightened environmental mindfulness, during which you perform many acts of kindness to create positive change.
- Forge alliances with kindred spirits to lead collective endeavours in pro-environmental.

The gift of time for a greener world

- Offer your time to environmental causes and community initiatives devoted to sustainability, enriching individual lives.
- Help neighbours embark on eco-friendly endeavours by engaging everyone in planet-friendly behaviours.
- Share your knowledge and skills through educational forums and workshops.

Performing several acts of kindness within a single day offers added benefits to our wellbeing (Lyubomirsky et al., 2005). Furthermore, the ethos of the Gift of Time (Fredrickson, 2009) advocates for sustained volunteerism in support of environmental causes, thereby highlighting a commitment to the wellbeing of individuals and the planet. By engaging with this pro-environmental kindness, we contribute to community collaboration, and sustainable existence.

Crafting your ecological expressive writing journey

To embark on your expressive writing journey for environmental wellbeing:

1 Grab a piece of paper and a pen.
2 Set an alarm for 20 minutes to delve into your deepest thoughts and feelings regarding biodiversity, pro-environmental actions, and the ecological impact on your life.
3 Explore your emotions and thoughts, encompassing your relationships with nature, experiences with environmental efforts, reflections on past eco-actions, and aspirations for a sustainable future.

This activity is most effective when repeated daily over the next 3–5 days.

Tips for an optimal eco-expressive writing experience

- **Duration Matters:** Even a single writing session can considerably impact (Greenberg et al., 1996). However, most studies recommend expressive writing for 20 minutes over 3–5 consecutive days.
- **Handwriting vs. Typing:** Opt for the traditional handwriting approach, which has proven more effective than typing (Brewin & Lennard, 1999).
- **Language Choice:** Bilingual individuals are encouraged to write in either their native language or a mix of both, as reported benefits are higher for those expressing themselves in two languages (Youngsuk, 2008).

Embrace the power of expressive writing as a tool for ecological expression, providing a means to navigate environmental challenges, discover eco-insights, and foster your emotional wellbeing intertwined with a commitment to biodiversity and pro-environmental actions.

Awe

Awe is a complex emotional phenomenon comprising several elements (Keltner & Haidt, 2003). Firstly, Awe typically arises from stimuli perceived as vast, grand, or beyond ordinary comprehension. This could include natural wonders like mountains or starry skies or human-made achievements like architectural marvels or artistic masterpieces. Secondly, Awe often occurs when individuals encounter something challenging their cognitive frameworks or beliefs. This need for accommodation can lead to cognitive expansion or openness to new ideas and perspectives.

It is a powerful emotional state characterised by intense positive emotions such as wonder, amazement, and reverence, often accompanied by a sense of insignificance or humility in the face of something greater than oneself. It often elicits feelings of connectedness to something larger or more significant than oneself, whether it be other people, nature, the universe, or a higher power. This sense of connection can foster feelings of unity, empathy, and compassion. Awe has been described as a self-transcendent emotion, temporarily shifting attention away from the self and towards something more significant. This can lead to diminished personal concerns or ego and increased focus on collective or existential matters.

Finally, Awe experiences have been associated with a sense of time dilation, where individuals may feel like time has slowed down or expanded. This altered perception of time can contribute to Awe's immersive and transformative nature. Overall, the psychological components of Awe contribute to its profound and multifaceted nature, making it an affluent area of study within psychology and neuroscience.

Research suggests that individuals who experience Awe in response to natural landscapes are more inclined to support environmental protection and conservation behaviours (Piff et al., 2015). This indicates that we can cultivate a connection to nature and appreciate its intrinsic value, motivating actions to preserve biodiversity. Similarly, Awe experienced in natural settings correlates with heightened concern for environmental issues and backing for conservation efforts (Bai et al., 2017). Consequently, those reporting greater awe in response to nature tend to engage more in pro-environmental behaviours like recycling and supporting biodiversity protection policies.

Furthermore, beyond individual actions, Awe can become a catalyst for broader societal shifts towards sustainability. Rudd et al. (2012) found that experiencing Awe in natural environments fosters feelings of interconnectedness with others and nature. This sense of interconnectedness can instil a collective responsibility for biodiversity conservation and environmental sustainability efforts.

Life satisfaction

Life satisfaction represents a cognitive facet of emotional wellbeing (Diener, 1984). It entails individuals reflecting on their lives, contemplating their past dreams, aspirations, and hopes, and assessing them in their current circumstances. If individuals find that their current circumstances fall short of the expectations they held for their lives, their satisfaction with life may decline. Conversely, if their achievements align with or exceed their initial hopes and aspirations, their sense of life satisfaction is likely to be heightened. However, the complexity of this concept is compounded by various factors, including the emotional state of individuals at the time of responding to such inquiries.

According to Kahnemann and Riis (2005), the emotional state of individuals significantly influences their responses to questions regarding life satisfaction. Their responses may vary depending on whether individuals are experiencing positive or negative emotions when reflecting on their lives. Thus, while life satisfaction is primarily viewed as a cognitive aspect of emotional wellbeing, it is inherently intertwined with emotions, blurring the boundaries between cognitive and affective dimensions.

This intricate interplay between cognition and emotion highlights the multifaceted nature of life satisfaction. By acknowledging the intertwined nature of cognitive and emotional aspects, we can better understand life satisfaction and its role in shaping emotional wellbeing.

A discernible relationship exists between life satisfaction and various pro-environmental behaviours. For instance, pro-environmental behaviour, such as recycling, water conservation, and purchasing "environmentally-friendly" products, has been identified as predictors of life satisfaction (Welsch & Kühling, 2010). Moreover, a comprehensive study involving over 2,000 individuals revealed that engaging in any of 37 pro-environmental actions significantly predicted participants' overall life satisfaction (Schmitt et al., 2018). Interestingly, of all the pro-environmental behaviours, those that involved social interactions emerged as particularly impactful on individuals' wellbeing. This highlights the importance of community engagement and collective action in fostering a sense of fulfilment and satisfaction in individuals' lives.

Furthermore, an intriguing nuance arises concerning participants' evaluations of their behaviour's sustainability. It's not merely the act of engaging in pro-environmental behaviour that contributes to increased life satisfaction but rather the perceived environmental impact of such actions (Binder & Blankenberg, 2017). Participants' assessments of the sustainability of their behaviour played a crucial role in shaping their emotional wellbeing, emphasising the significance of perceived efficacy and positive outcomes in environmental stewardship efforts.

Thus, emotional wellbeing is intricately intertwined with pro-environmental actions, with individuals' sense of fulfilment and satisfaction influenced by both the nature of their environmental behaviours and their perceived impact on the world around them. This highlights the potential of pro-environmental actions to benefit the planet and enhance individuals' subjective wellbeing all at the same time.

Conclusion

Th chapter argued that engaging in pro-environmental actions could enhance participants' emotional wellbeing. Our discussion has been grounded in the Broaden-and-Build theory, which highlights the beneficial effects of positive emotions and their capacity to undo negative emotions, leading to an upward spiral of emotional growth. Through our exploration, we have delved into various emotions, such as gratitude and awe, demonstrating their role in fostering emotional resilience and wellbeing.

Furthermore, we have examined emotional wellbeing from the life satisfaction perspective, recognising it as a cognitive component intricately linked with emotional wellbeing. Our review has shown a significant correlation between pro environmental actions and life satisfaction, suggesting that individuals who engage in such behaviours experience higher overall satisfaction with their lives.

While the evidence supports the notion that participating in pro-environmental actions can indeed contribute to individuals' emotional wellbeing, it is essential to acknowledge the complexities inherent in this relationship. Future research endeavours must delve deeper into these intricacies to better understand how different environmental actions impact emotional wellbeing.

From a positive psychology standpoint, it is encouraging to consider the potential for meaningful engagement in environmental improvement to serve as a positive psychology intervention. By viewing environmental stewardship as a pathway to personal growth and wellbeing, individuals can proactively enhance their emotional health while contributing to broader environmental goals.

In the forthcoming chapter, we will delve into other psychological elements of wellbeing, broadening our understanding of the multifaceted interplay between human psychology and environmental sustainability. Through this exploration, we aim to uncover additional insights into how psychological factors shape individuals' attitudes, behaviours, and perceptions regarding environmental issues.

References

Algoe, S. B., Gable, S. L., & Maisel, N. C. (2010). It's the little things: Everyday grattitude as a booster shot for romantic relationships. *Personal Relationships*, 17(2), 217–233.

Algoe, S. B., Haidt, J., & Gable, S. L. (2008). Beyond reciprocity: Gratitude and relationships in everyday life. *Emotion (Washington, D.C.)*, 8(3), 425–429. https://doi.org/10.1037/1528-3542.8.3.425

Bahrami, F., Kasaei, R., & Zamani, A. (2012). Preventing worry and rumination by induced positive emotion. *International Journal of Preventive Medicine*, 3(2), 102–109.

Bai, Y., et al. (2017). Awe, the diminished self, and collective engagement: Universals and cultural variations in the small self. *Journal of Personality and Social Psychology*, 113(2), 185–209. https://doi.org/10.1037/pspa0000087

Bain, P., et al. (2016). Co-benefits of addressing climate change can motivate action around the world. *Nature Climate Change*, 6, 154–157. https://doi.org/10.1038/nclimate2814

Bartlett, M. Y., & De Steno, D. (2006). Gratitude and prosocial behaviour. *Psychological Science*, 17(4), 319–325.

Berkes, F., & Ross, H. (2013). Community resilience: Toward an integrated approach. *Society & Natural Resources*, 26(1), 5–20. https://doi.org/10.1080/08941920.2012.736605

Binder, M., & Blankenberg, A.-K. (2017). Green lifestyles and subjective well-being: More about self-image than actual behavior? *Journal of Economic Behavior & Organization*, 137, 304–323. https://doi.org/10.1016/j.jebo.2017.03.009

Brewin, C. R., & Lennard, H. (1999). Effects of mode of writing on emotional narratives. *Journal of Traumatic Stress*, 12(2), 355–361. https://doi.org/10.1023/A:1024736828322

Burton, C. M., & King, L. A. (2004). The health benefits of writing about intensely positive experiences. *Journal of Research in Personality*, 38(2), 150–163. https://doi.org/10.1016/S0092-6566(03)00058-8

Carter, D. M. (2011). Recognizing the role of positive emotions in fostering environmentally responsible behaviors. *Ecopsychology*, 3(1), 1–5. https://doi.org/10.1089/eco.2010.0071

Chen, L., Liu, J., Fu, L., Guo, C., & Chen, Y. (2022). The impact of gratitude on connection with nature: The mediating role of positive emotions of self-transcendence. *Frontiers in Psychology*, 13, 908138. https://doi.org/10.3389/fpsyg.2022.908138

Curry, O. S., Rowland, L. A., Van Lissa, C. J., Zlotowitz, S., McAlaney, J., & Whitehouse, H. (2018). Happy to help? A systematic review and meta-analysis of the effects of performing acts of kindness on the well-being of the actor. *Journal of Experimental Social Psychology*, 76, 320–329.

Dalebroux, A., Goldstein, T. R., & Winner, E. (2008). Short-term mood repair through art-making: Positive emotion is more effective than venting. *Motivation and Emotion*, 32(4), 288–295. https://doi.org/10.1007/s11031-008-9105-1

Davidson, R. J., & Beagley, S. (2012). *The emotional life of your brain: How its unique patterns affect the way you think, feel, and live--And how you can change them.* Avery Publishing Group.

DeShea, L. (2003) A scenario based scale of willingness to forgive. *Individual Differences Research*, 1, 201–217.

Diener, E. (1984). Subjective well-being. *Psychological Bulletin*, 95, 542–575.

Diener, E. (2010). New well-being measures: Short scales to assess flourishing and positive and negative feelings. *Social Indicators Research*, 97, 143–156.

Dietz, T., Fitzgerald, A., & Shwom, R. (2005). Environmental values. *Annual Review of Environmental Resources*, 30, 335–372.

Digdon, N., & Koble, A. (2011). Effects of constructive worry, imagery distraction, and gratitude interventions on sleep quality: A Pilot trial. *Applied Psychology: Health & Well-being*, 3(2), 193–206.

Dodds, J. (2021). The psychology of climate anxiety. *BJPsych Bulletin*, 45(4), 222–226. https://doi.org/10.1192/bjb.2021.18

Emmons, R. A., & McCullough, M. E. (2003). Counting blessings versus burdens: An experimental investigation of gratitude and subjective well-being in daily life. *Journal of Personality and Social Psychology*, 84(2), 377–389. https://doi.org/10.1037/0022-3514.84.2.377

Emmons, R. A., & Shelton, C. M. (2005). Gratitude and the science of positive psychology. In C. R. Snyder & S. J. Lopez (Eds.), *Handbook of positive psychology* (pp. 459–471). Oxford University Press.

Fredrickson, B. L. (2001). The role of positive emotions in positive psychology: The broaden-and-build theory of positive emotions. *American Psychologist*, 56(3), 218–226. https://doi.org/10.1037/0003-066X.56.3.218

Fredrickson, B. L. (2004). Gratitude, like other positive emotions, broadens and builds. In R. A. Emmons & M. E. McCullough (Eds.), *The psychology of gratitude* (pp. 145–166). Oxford University Press.

Fredrickson, B. L. (2009). *Positivity: Groundbreaking research reveals how to embrace the hidden strength of positive emotions, overcome negativity, and thrive.* Crown Archetype.

Fredrickson, B. L., Tugade, M. M., Waugh, C. E., & Larkin, G. R. (2003). What good are positive emotions in crises?: A prospective study of resilience and emotions following the terrorist attacks on the United States on September 11, 2001. *Journal of Personality and Social Psychology*, 84, 377–389.

Greenberg, M. A., Wortman, C. B., & Stone, A. A. (1996). Emotional expression and physical health: Revising traumatic memories or fostering self-regulation? *Journal of Personality and Social Psychology*, 71(3), 588–602. https://doi.org/10.1037/0022-3514.71.3.588

Hefferon, K., & Boniwell, I. (2011). *Positive psychology: Theory, research and applications*. McGraw-Hill.

Kahneman, D., & Riis, J. (2005). Living, and thinking about it: Two perspectives on life (Link downloads document). In F. A. Huppert, N. Baylis, & B. Keverne (Eds.), *The science of well-being* (pp. 285–304). Oxford University Press.

Keltner, D., & Haidt, J. (2003). Approaching awe, a moral, spiritual, and aesthetic emotion. *Cognition and Emotion*, 17(2), 297–314. https://doi.org/10.1080/02699930302297

Kuhbandner, C., Lichtenfeld, S., & Pekrun, R. (2011). Always look on the broad side of life: Happiness increases the breadth of sensory memory. *Emotion*, 11(4), 958–964. https://doi.org/10.1037/a0024075

Lepore, S. J., & Revenson, T. A. (2006). Resilience and posttraumatic growth: Recovery, resistance, and reconfiguration. In L. G. Calhoun & R. G. Tedeschi (Eds.), *Handbook of posttraumatic growth: Research & practice* (pp. 24–46). Lawrence Erlbaum Associates Publishers.

Lutz, P. K., Zelenski, J. M., & Newman, D. B. (2023). Eco-anxiety in daily life: Relationships with well-being and pro-environmental behavior. *Current Research in Ecological and Social Psychology*, 4. https://doi.org/10.1037/1528-3542.8.3.425

Lyubomirsky, S. (2007). *The how of happiness: A practical guide to getting the life you want*. Sphere.

Lyubomirsky, S., Sheldon, K. M., & Schkade, D. (2005). Pursuing happiness: The architecture of sustainable change. *Review of General Psychology*, 9, 111–131.

Masten, A. S., Best, K. M., & Garmezy, N. (1990). Resilience and development: Contributions from the study of children who overcome adversity. *Development and Psychopathology*, 2(4), 425–444. https://doi.org/10.1017/S0954579400005812

Mauss, I. B., Savino, N. S., Anderson, C. L., Weisbuch, M., Tamir, M., & Laudenslager, M. L. (2012). The pursuit of happiness can be lonely. *Emotion (Washington, D.C.)*, 12(5), 908–912. https://doi.org/10.1037/a0025299

Mauss, I. B., Tamir, M., Anderson, C. L., & Savino, N. S. (2011). Can seeking happiness make people unhappy? [corrected] Paradoxical effects of valuing happiness. *Emotion (Washington, D.C.)*, 11(4), 807–815. https://doi.org/10.1037/a0022010

McCullough, M. E., Emmons, R. A., & Tsang, J.-A. (2002). The grateful disposition: A conceptual and empirical topography. *Journal of Personality and Social Psychology*, 82(1), 112–127. https://doi.org/10.1037/0022-3514.82.1.112

Miller, D. (1995). Citizenship and pluralism. *Political Studies*, 43(3), 432–450. https://doi.org/10.1111/J.1467-9248.1995.TB00313.X

Piff, P. K., Dietze, P., Feinberg, M., Stancato, D. M., & Keltner, D. (2015). Awe, the small self, and prosocial behavior. *Journal of Personality and Social Psychology*, 108(6), 883–899. https://doi.org/10.1037/pspi0000018

Rowe, G., Hirsh, J. B., & Anderson, A. K. (2007). Positive affect increases the breadth of attentional selection. *Proceedings of the National Academy of Sciences of the United States of America*, 104(1), 383–388. https://doi.org/10.1073/pnas.0605198104

Rudd, M., Vohs, K. D., & Aaker, J. (2012). Awe expands people's perception of time, alters decision making, and enhances well-being. *Psychological Science*, 23(10), 1130–1136. https://doi.org/10.1177/0956797612438731

Schmitt, M. T., Aknin, L. B., Axsen, J., & Shwom, R. L. (2018). Unpacking the relationships between pro-environmental behavior, life satisfaction, and perceived ecological threat. *Ecological Economics*, 143, 130–140. https://doi.org/10.1016/j.ecolecon.2017.07.007

Seligman, M. E. P. (2003). Positive psychology: Fundamental assumptions [Editorial]. *The Psychologist, 16*(3), 126–127.

Seligman, M. E. P., Steen, T. A., Park, N., & Peterson, C. (2005). Positive psychology progress: Empirical validation of Interventions. *American Psychological Association*, 60, 410–421.

Stoverink, A. C., Kirkman, B. L., Mistry, S., & Rosen, B. (2020). Bouncing back together: Toward rhetorical model of work team resilience. *The Academy of Management Review*, 45(2), 395–422. https://doi.org/10.5465/amr.2017.0005

Sun, J., Ma, B., & Wei, S. (2023). Same gratitude, different pro-environmental behaviors? Effect of the dual-path influence mechanism of gratitude on pro-environmental behavior. *Journal of Cleaner Production*, 415. https://doi.org/10.1016/j.jclepro.2023.137779

Tam, K.-P. (2022). Gratitude to nature: Presenting a theory of its conceptualization, measurement, and effects on pro-environmental behavior. *Journal of Environmental Psychology*, 79, 1–14. https://doi.org/10.1016/j.jenvp.2021.101754

Watkins, P. C. (2004). Gratitude and subjective well-being. In R. A. Emmons & M. E. McCullough (Eds.), *The psychology of gratitude* (pp. 167–192). Oxford University Press.

Watkins, P. C., Scheer, J., Ovnicek, M., & Kolts, R. (2006). The debt of gratitude: Dissociating gratitude and indeptness. *Cognition and Emotion*, 20, 217–241.

Welsch, H., & Kühling, J. (2010). Pro-environmental behavior and rational consumer choice: Evidence from surveys of life satisfaction. *Journal of Economic Psychology*, 31(3), 405–420. https://doi.org/10.1016/j.joep.2010.01.009

Wood, A. M., Froh, J., & Geraghty, A. (2010). Gratitude and well-being: A review and theoretical integration. *Clinical Psychology Review*, 30(7), 890–905.

Wood, A. M., Joseph, S., Lloyd, J., & Atkins, S. (2009). Gratitude influences sleep through the mechanism of pre-sleep cognitions. *Journal of Psychosomatic Research*, 66(1), 43–48.

Youngsuk, C. (2008). *Politicizing Asian American literature towards a critical multiculturalism*. Routledge.

Zelenski, J. M., & Desrochers, J. E. (2021). Can positive and self-transcendent emotions promote pro-environmental behavior? *Current Opinion in Psychology*, 42, 31–35. https://doi.org/10.1016/j.copsyc.2021.02.009

6 Individuals' psychological wellbeing

Paul's native woodland, which he planted two decades ago, stands at his doorstep. Paul embarked on his journey of woodland planting out of necessity. Facing a small plot of land with little interest in farming, Paul and his brother researched alternatives, ultimately opting for a pilot project to establish a native woodland. The process involved careful selection of native tree species, preparing the soil, and planting the saplings. The project initially covered planting costs and provided subsidies for several years, although these subsidies have since ceased.

Considerations such as biodiversity and carbon capture should have been discussed when planting, reflecting a different era of environmental awareness. Despite this, Paul feels a profound sense of accomplishment and pride in his woodland, given its proximity to his family home. He views the decision to plant the woodland as a moral imperative, although familiarity sometimes breeds complacency, leading to occasional lapses in appreciation.

Paul took courses in biodiversity, which deepened his appreciation for the flora and fauna inhabiting the woodland. He likened this newfound knowledge to knowing someone's name in a social setting – recognising a plant or bird in the forest allowed him to relate to it, fostering a deeper connection. Paul's family also learned which plants in their woodland were edible, enhancing meals with fresh forest finds like berries and mushrooms, and their distinct aromas. This added variety to their diet and brought them closer to nature.

Hosting visitors, such as an active retirement group, allowed Paul to share the sensory experiences and nostalgic memories evoked by the woodland. His enjoyment extended to hosting local schools, with younger visitors often declaring their woodland excursion the "best day of their life". Recalling birthday parties filled with laughter and joyous chaos in nature, Paul found fulfilment in witnessing both children and adults revel in the wonders of the woodland. The woodland, besides its environmental benefits, has become a source of joy and community bonding.

DOI: 10.4324/9781003452676-8

Individuals from farming backgrounds often feel a strong inclination towards agricultural activities, driven by a desire to produce and contribute. Paul, for instance, believed there was untapped potential in his woodland, envisioning it as a nature school or a destination for school tours. However, he grappled with the idea, recognising the emotional complexity of sharing the woodland with others. Balancing visitor access with biodiversity preservation proved challenging, as excessive foot traffic could harm the ecosystem and necessitate litter cleanup. These challenges, however, did not deter Paul's passion for his woodland.

Despite the stress of managing a diverse habitat and confronting issues like Ash dieback, which threatens ash trees, Paul remained passionate about showcasing the woodland to local school children. However, the lack of financial subsidies for established woodlands added to the strain, impacting families reliant on them for survival. The emotional and financial strains were significant, but Paul's engagement and dedication to the woodland and its educational value remained unwavering.

Psychological wellbeing encompasses more than subjective or emotional wellness, delving into dimensions beyond mere contentment. While the foundational model of psychological wellbeing was introduced by Ryff in the 1980s (Ryff, 1989), the emergence of the positive psychology movement and advancements in positive health research have broadened this perspective. These developments have incorporated additional concepts that transcend emotional wellbeing, embracing eudaimonic aspects of wellness aimed at facilitating individuals in living fulfilling lives (Huppert, 2009).

In this chapter, we explore selected concepts relevant to psychological wellbeing, particularly those that demonstrate positive impacts resulting from pro-environmental actions. We aim to uncover concepts that benefit from participation in environmental stewardship, potentially leading to reversing negative environmental trends and enhancing overall human wellbeing.

We endeavour to shed light on the reciprocal relationship between individual wellbeing and environmental engagement by delving into these interconnected realms of psychology and environmentalism. Through this exploration, we seek to explore how nurturing our natural surroundings can contribute not only to the health of the planet but also to the flourishing of human minds and spirits.

Limited research explores the intersection of overall psychological wellbeing and its relationship with pro-environmental behaviours. While numerous studies investigate the various components of psychological wellbeing, this chapter aims to delve into these facets more comprehensively.

Emerging research suggests a notable correlation between psychological wellbeing and engagement in pro-environmental behaviours. Notably, Corral-Verdugo and colleagues (2013) have presented evidence demonstrating that individuals who actively participate in sustainable practices tend to exhibit higher psychological wellbeing levels than their counterparts who do not. Nevertheless, the intricacies of this connection warrant further examination. Recent studies, such as that by Liu et al. (2022), have begun to explore the nuances of this relationship. Their findings indicate that factors such as the amount of time spent in nature and the level of contact individuals have with natural environments can significantly impact the association between psychological wellbeing and pro-environmental behaviours.

While existing studies have shed some light on a relationship between psychological wellbeing and pro-environmental behaviours, the broader realm of psychological wellbeing in relation to these behaviours remains largely unexplored. Therefore, ongoing research into the underlying mechanisms and contextual factors influencing this relationship is not only necessary but also holds the potential for a more comprehensive understanding.

Engagement

In our fast-paced modern lives, experiencing boredom may seem like an anomaly, a rare occurrence amidst the constant buzz of activity. Yet, when boredom strikes it often triggers a reflexive reach for our phones – an attempt to fend off monotony with passive distractions or digital connections. In this world of perpetual connectivity, two distinct forms of boredom persist, each leaving its mark on our mental landscape and influencing our wellbeing.

At the heart of our exploration lies state boredom – the fleeting feelings of boredom – and trait boredom – a predisposition towards this state (Fahlman et al., 2013). For those who frequently battle boredom, the challenge of engagement looms large, as engagement serves as the key to motivation. To understand its significance, we delve into the essence of Flow – a state of profound absorption and unwavering focus when pursuing an activity (Csikszentmihalyi, 1997). Flow immerses individuals in a state of purpose, where time fades away and the outside world becomes insignificant, leaving only the task at hand. However, paradoxically, the very awareness of the flow state can disrupt its delicate balance, breaking the flow experience.

Statistics reveal that roughly one-tenth to one-sixth of people experience Flow in their lives, while a similar fraction have never experienced it (Hefferon & Boniwell, 2011). This disparity can be attributed to two main factors. Firstly, the presence of an autotelic personality, characterised by boundless curiosity, relentless persistence, and intrinsic motivation to transform boredom into interest (Tse et al., 2020). Thus, individuals with autotelic personality are more likely to experience Flow states. Secondly, research and practice on how to experience Flow every day is scarce making it all the more fascinating to explore.

Flow addresses the fundamental human desires for competence and autonomy, paving the way towards enhanced wellbeing (Ilies et al., 2017). Furthermore, it heralds increased engagement and productivity, surpassing individual pursuits to enriching collective endeavours (Fullagar & Fave, 2017). Indeed, throughout history, the most creative minds have immersed themselves in Flow states, attesting to its transformative power and role in unlocking human potential (Csikszentmihalyi, 1997).

As we delve deeper into the concept of Flow, it's crucial to recognise its profound relevance to environmental stewardship and biodiversity conservation. Despite its potential to serve as a catalyst for engaging in pro-environmental actions or deriving psychological wellbeing from this engagement, research in this area remains scarce, which is somewhat surprising considering the transformative power that such experiences could hold for many individuals, sparking a new level of environmental consciousness and action.

Recent studies indicate that the state of Flow derived from outdoor recreational activities has a significant impact on pro-environmental behavioural intentions (Han, 2023). This not only suggests that Flow experiences in natural settings can inspire individuals to adopt more environmentally conscious behaviours, but also holds the promise of a future where environmental stewardship is a natural extension of our daily lives.

While Flow has traditionally been associated with domains such as work, sports, and creativity, its principles have the capacity to extend beyond these realms, guiding individuals towards environmental activism. In the context of environmental engagement, Flow could manifest in various forms, intertwining with pro-environmental behaviours and efforts towards biodiversity preservation. This highlights the potential of each individual to make a significant impact in the realm of environmental conservation.

Consider, for instance, the conservationists working in a community park, where each action contributes to the renewal of the local environment. Similarly, the wildlife enthusiast, captivated by nature, becomes a witness to intricate changes in biodiversity. Each individual embodies Flow, driven by purpose and a profound engagement. As such, Flow experiences have the potential to serve as a gateway to environmental activism, fostering a connection between individuals and the natural world and inspiring meaningful action towards conservation and sustainability.

Sense of accomplishment

Sense of accomplishment, the emotional satisfaction derived from completing tasks or achieving goals, plays a crucial role in individuals' psychological wellbeing (Locke & Latham, 2002) in the context of individual and community wellbeing. This feeling of fulfilment and pride arises from competence and self-efficacy, whether from conquering small tasks or significant life milestones (Sheldon & Elliot, 1999).

Achieving tasks and reaching goals can significantly enhance individuals' self-esteem and self-confidence (Bandura, 1977). Successes contribute to a positive self-image and reinforce beliefs in one's abilities and competencies. Moreover, experiencing a sense of accomplishment can fuel intrinsic motivation, inspiring individuals to set and pursue increasingly ambitious objectives (Sheldon & Elliot, 1999).

The psychological benefits of a sense of accomplishment extend beyond mere satisfaction, encompassing the reduction of stress and anxiety, which are closely linked to the concept of eco-anxiety. When individuals experience a sense of achievement in completing tasks or achieving goals related to environmental stewardship, they often find relief from the burdens of stress and anxiety.

Research suggests that accomplishing tasks gives individuals a heightened sense of control and mastery over their environment, contributing to reduction in stress and fostering resilience in managing challenges (Locke & Latham, 2002). This sense of control and mastery is particularly relevant in environmental concerns, where feelings of eco-anxiety can arise from a perceived lack of agency in addressing complex environmental issues.

Individuals can regain a sense of agency and efficacy by engaging in pro-environmental actions and witnessing tangible outcomes, thereby mitigating eco-anxiety and promoting psychological wellbeing. This highlights the importance of recognising the therapeutic value of environmental engagement and its positive impact on mental health and resilience.

Furthermore, a sense of accomplishment is closely linked to happiness, fulfilment, and life satisfaction (Sheldon & Elliot, 1999). Achieving goals contributes to overall psychological wellbeing and gives individuals a sense of purpose and meaning.

In the context of pro-environmental actions, a sense of accomplishment can be leveraged to inspire and sustain engagement in sustainable behaviours. Here are some of the activities that can help individuals feel like they have accomplished something significant in the context of pro-environmental actions:

- Setting achievable environmental goals, such as reducing personal carbon footprint or participating in recycling programmess, can evoke a sense of accomplishment.
- Recognising and celebrating milestones in pro-environmental actions reinforces a sense of accomplishment and motivates continued engagement.
- Publicly acknowledging achievements, such as reaching a community recycling target or successfully advocating for local environmental policies, can inspire others to participate.

- Offering feedback and recognition for environmentally responsible behaviours enhances individuals' sense of accomplishment and reinforces the value of pro-environmental actions.
- Positive reinforcement, such as acknowledging efforts to conserve resources or protect ecosystems, encourages sustained engagement.
- Engaging in collaborative environmental initiatives allows individuals to share the sense of accomplishment that comes from collective achievement.
- Working together towards shared environmental goals enhances social connectedness and amplifies the impact of individual efforts.

In conclusion, a sense of accomplishment is a powerful motivator for positive environmental change. By harnessing individuals' innate desire for achievement and recognition, pro-environmental initiatives can inspire action and promote sustainable behaviours that contribute to the wellbeing of both people and the planet. A sense of accomplishment can drive meaningful progress towards environmental sustainability through goal-setting, recognition, collaboration, and celebration of achievements.

Anticipation

As Homo prospectus, humans are inherently future-oriented beings driven by the innate instinct of survival (Seligman et al., 2016). While past adversities undoubtedly shape our present wellbeing, it is essential to acknowledge the significant impact of future-oriented concerns on our daily lives. The fear of tomorrow's environmental challenges can often dampen our motivation to take action today, impeding our collective efforts for a sustainable future. However, the concept of anticipation offers a powerful lens through which we can navigate our environmental engagement, especially in the context of community we want to motivate to action.

Anticipation is a multifaceted cognitive process encompassing various psychological phenomena, including imagination, expectation, and emotional preparation. At its core, anticipation involves mentally simulating and preparing for future events, whether positive or negative. Depending on the anticipated outcome, this process often triggers a range of emotions, from excitement and eagerness to anxiety and apprehension. These emotional responses can significantly influence our behaviour and decision-making, shaping our actions and reactions in anticipation of future events.

Anticipation is pivotal in pro-environmental behaviour and action, shaping individuals' aspirations and motivations towards environmental stewardship. Through anticipation, individuals envision positive future outcomes, setting goals and fueling their commitment to sustainable practices. For instance, individuals eagerly anticipate the prospect of a cleaner environment, envisioning lush green landscapes, abundant wildlife, and cleaner air and waterways due to their collective efforts in environmental conservation (Bryant, 2003).

Future-focused environmental journaling (adapted from Roepke et al., 2018)

If you have recently faced environmental adversity or challenges, set aside 15 minutes once a week for the next month to engage in prospective writing about future opportunities related to pro-environmental behaviour and biodiversity conservation. Reflect on the lessons learned from your experiences and identify new doors that may open for you in the coming months. These opportunities involve exploring new ways to engage with nature, setting conservation goals, connecting with like-minded individuals or organisations, or making lifestyle changes to reduce your environmental footprint. While past losses and challenges may have been difficult, they have paved the way for new growth and possibilities in your journey towards environmental stewardship.

This cognitive process, facilitated by the human brain's prefrontal cortex, empowers individuals to plan and prepare for the preservation of our planet's natural resources (Melges, 1990). Individuals anticipate the potential benefits of their environmental actions, so they are motivated to take proactive steps towards sustainability and ecological restoration.

Individuals often exhibit an optimistic bias, overestimating the satisfaction they will derive from future environmental successes (Gilbert & Wilson, 2007). This optimistic outlook is a catalyst for community engagement, inspiring individuals to invest their time and resources in environmental initiatives with the anticipation of positive outcomes. By embracing anticipation and fostering a forward-thinking mindset, individuals and communities can unlock the full potential of their collective actions in environmental conservation. Anticipation enriches the present endeavours in environmental stewardship, serving as a guiding force towards a future of ecological harmony and resilience.

Through shared anticipation of a greener tomorrow, communities unite in their commitment to environmental sustainability, fostering a sense of belonging and collective purpose. Conversely, anticipating environmental challenges can mobilise communities to address pressing issues, channelling collective efforts towards finding solutions and mitigating potential environmental threats.

In summary, anticipation drives pro-environmental behaviour and action, motivating individuals and communities to work towards a sustainable and resilient future. By harnessing the power of anticipation, we can cultivate a sense of optimism and purpose in our environmental endeavours, paving the way for a brighter and more sustainable tomorrow.

Hope and optimism

Understanding the profound influence of optimism bias, especially in the realm of pro-environmental behaviour, is of paramount importance. Individuals naturally tend to exhibit an optimism bias (Sharot et al., 2007) when navigating various situations in life. This bias serves as a driving force, encouraging individuals to persist in their environmentally conscious pursuits while anticipating positive outcomes from adopting sustainable living practices. This optimistic outlook is prevalent as individuals envision a longer and healthier life, along with successfully addressing environmental challenges such as pollution and the impacts of climate change. Even when confronted with compelling facts that might logically challenge these optimistic beliefs, the brain adeptly counters them, adopting a positive illusion mindset, as explored by Sharot et al. (2011).

The inclination towards positive illusions becomes more pronounced with age, as highlighted in the research by Chowdhury et al. (2014). Individuals tend to become increasingly optimistic about the effectiveness of their actions as they age, thereby presenting a unique opportunity for environmental scientists to engage with this specific demographic. In sustainable living, the optimism bias emerges as a powerful driving force, fostering a sustained commitment to eco-friendly behaviours and instilling a positive perspective regarding the potential success of collective environmental actions.

Optimism is a belief in a positive outcome of our actions or the collective actions of others. When exploring optimism, we encounter two predominant research perspectives: one perceiving it as a personality trait, according to which some of us have a genetic predisposition to optimism, and the other framing it as a cognitive process available to everyone, regardless of their predispositions.

The latter viewpoint, epitomised by the work of psychologist Martin Seligman, provides compelling evidence that optimism is not just a trait but a skill that can be acquired and cultivated (Seligman, 1998). Seligman posits that optimism reflects how individuals interpret setbacks and victories. When confronted with a setback, those with an optimistic disposition tend to believe that:

Various factors shape their capacity to contribute to the planet's wellbeing. Those with this perspective don't attribute blame to themselves or others, even if they played a role in the challenges. Instead, they take a broader view, considering a multitude of factors influencing people's behaviours. Their attention is directed away from assigning blame and towards influencing a range of circumstances. They view the situation as temporary and believe in their ability to initiate change. For instance, they might persuade others about the necessity for change and compile a list of actions to alter circumstances or adopt different practices. This viewpoint remains specific to particular circumstances. Even if one attempt doesn't yield the desired results, they avoid catastrophising. They perceive it as a singular effort and maintain the belief that other elements of pro-environmental action or related circumstances can influence future endeavours to safeguard the planet.

Conversely, individuals with a pessimistic outlook are inclined to believe that these circumstances are often ascribed to them or other individuals, highlighting the enduring characteristics of everyone involved, such as human stupidity, greed, envy, and other facets of the human psyche seen as resistant to change. This viewpoint cultivates a feeling of helplessness when confronted with challenges, diminishing the inclination to initiate any change. Alternatively, their attempts may stem from a condescending perspective, reducing the likelihood of others listening. They view the setback as everlasting. They hold the belief that they lack the capacity to change the situation, which may result in feelings of helplessness and despair. The pervasive negative implications span all their endeavours, highlighting the urgent need to combat such pessimism for the sake of our environment.

Hence, both optimism and pessimism can be considered distortions or alterations of perception. The crucial questions are: Which mindset is more likely to inspire individuals towards pro-environmental action? Which attitude encourages individuals to resiliently persevere after facing failure? Which perspective has the potential to initiate positive climate change? Considering the potential for helplessness, despair, and inaction associated with pessimistic thinking, which may lead to anxiety and depression, the optimistic viewpoint emerges as a more adaptive approach to pro-environmental action. It should be advocated, particularly among activists.

The key concept here highlights that our feelings of optimism or pessimism are not solely determined by external events but rather by the habitual thought processes we employ. Importantly, this cognitive process is not fixed; it is malleable and can be intentionally changed over time. This insight offers a powerful tool for shaping more optimistic perspectives, especially in the context of pro-environmental actions, empowering individuals to take control of their outlook and feel hopeful and motivated.

It's crucial to emphasise that optimism and pessimism can manifest in various contexts, extending from individuals' personal lives (for instance, a pessimistic belief that their small contribution won't alleviate climate devastation), collective groups like families or societies (such as a pessimistic belief that the government doesn't care), to broader global phenomena (like the pessimistic belief that environmental damage is irreversible). All these pessimistic thoughts could ultimately cause inaction. Consequently, an individual showcasing optimism in their personal life might adopt a pessimistic outlook regarding climate change. Conversely, someone with a pessimistic disposition may have an optimistic perspective on global issues. This highlights the need to examine these concepts comprehensively within all relevant contexts.

ABCD tool for optimistic thinking (adapted from Seligman, 1998)

Embark on a week-long journey of introspection, dedicating a moment each day to contemplate a challenging or stressful situation in the context of climate change, sustainability, biodiversity, and pro-environmental action:

Adversity (A)

Describe the challenging situation objectively, focusing on what caused it. Examine its implications for the broader context of environmental concerns.

Beliefs (B)

Write down your interpretation of the event and how you explain it to yourself, mindful of its potential impact on climate change, sustainability, or biodiversity. Consider the cognitive aspects that shape your pro-environmental attitudes.

Consequences (C)

Document your emotional responses and subsequent actions triggered by the identified beliefs. Reflect on the broader implications for pro-environmental behaviour, recognising the interconnected nature of emotions and environmental commitment.

Upon acknowledging your beliefs adopt strategies conducive to thinking optimistically in the environmental context.

Distract

Redirect your focus through engaging activities. Use techniques such as saying STOP and tapping the table, snapping your wrist with a rubber band, or promptly jotting down pessimistic thoughts about climate change or environmental challenges.

Distance

Reaffirm that your beliefs are subjective interpretations. Acknowledge that a pessimistic viewpoint is merely one set of beliefs coexisting with alternative, more optimistic explanations that can be explored within the environmental narrative.

Dispute

Question the evidence supporting your beliefs and evaluate its alignment with the reality of environmental challenges. Delve into alternative, optimistic perspectives concerning climate change and sustainability. Assess the implications of your beliefs, discerning which set is most conducive to improving your mood and contributing to pro-environmental goals.

Carr (2020) presents practical approaches to enrich the ABCD identification process, which could be tailored for pro-environmental initiatives. A notably effective strategy is Socratic Questioning, offering a valuable tool to challenge pessimistic beliefs. Some questions encompassed in this method include:

- Is there any evidence contradicting my belief?
- Are there relevant facts about the situation that I am neglecting?
- What inquiries would I direct to my closest friend if they shared a similar viewpoint?
- Am I unfairly attributing blame to myself for factors beyond my control?

Another impactful technique involves engaging in behavioural experiments to generate evidence evaluating pessimistic beliefs. This technique has a significant positive impact, as it allows individuals to perceive failure in a specific eco-friendly task due to feelings of incompetence and then embark on a moderately challenging environmental undertaking, acknowledging their proficiency in the process. This practice effectively facilitates the development of optimism for future success in such tasks.

A growing body of research explores the connection between pro-environmental actions and optimism, revealing insights into how positive messaging can influence behaviours and attitudes towards environmental issues. For instance, recent studies have demonstrated that optimistic messaging regarding environmental challenges can lead to improved pro-environmental behaviour, such as recycling, reducing energy consumption, and supporting renewable energy initiatives. It can even foster support for innovative solutions like geoengineering (MacKinnon et al., 2022).

This finding aligns with broader psychological theories, indicating that people are more inclined to follow optimistic leaders and adopt optimistic perspectives. Indeed, an examination of speeches delivered by U.S. presidents supports this notion, revealing a predominance of optimistic viewpoints among the elected leaders.

For example, President Ronald Reagan was often noted for his optimistic rhetoric. In his inaugural address in 1981, he famously declared, "In this present crisis, government is not the solution to our problem; government is the problem." This statement reflects an optimistic belief in the potential of the American people and the private sector to address challenges and drive progress.

Similarly, President Obama frequently emphasised the importance of hope and optimism in his speeches, famously using the slogan "Yes We Can" during his 2008 presidential campaign. In his farewell address 2017, he stated, "I am asking you to believe. Not in my ability to bring about change – but in yours." These examples highlight the power of optimism in leadership, inspiring and motivating others to take action.

The link between optimism and pro-environmental actions emphasises the importance of framing environmental discourse in a positive light. Optimistic narratives inspire hope and confidence in individuals by highlighting the potential for positive outcomes and opportunities for collective action, motivating them to actively engage in behaviours that contribute to environmental sustainability. This optimistic messaging has the potential to catalyse meaningful change in environmental attitudes and behaviours.

Moreover, understanding the interplay between optimism and pro-environmental actions can inform the development of more effective communication strategies and policy interventions aimed at promoting environmental stewardship. By harnessing the power of optimism, while at the same time avoiding unrealistic optimism, policymakers, activists, and communicators can empower the public, fostering a shared commitment to building a more sustainable future. It's your actions that can make a difference.

Hopefulness

The extensive literature on hope highlights its profound significance and delves into its intricate nature, as evidenced by numerous works dedicated to its exploration. However, research takes a unique perspective by focusing on the role of hope in pro-environmental action. With over 30 theories or definitions and two primary perspectives, hope can be perceived either as an emotion or a thinking process culminating in the feeling of hope.

> In a National Geographic blog dated 22 May 2020, titled "30 Days of Hope", a project aimed at facilitating coexistence between elephants and humans in Vietnam was highlighted. The Quang Nam Elephant Reserve faced the challenge of preventing elephants from straying into gardens and agricultural areas in search of food, which posed risks to both the human population and the elephants.
>
> To address this issue, staff at the elephant reserve implemented a solution involving planting nearly 2 kilometres of a native tree species called Bo Ket, known for its sharp thorns. These quickly growing trees were a natural barrier to deter elephants from leaving the reserve.

Additionally, the local community could use the leaves of the Bo Ket tree in traditional medicine and cosmetic products, providing an additional source of income for residents.

By taking measures to ensure the safety of elephants and eliminate the risk of wildlife-human conflicts, the community demonstrated kindness towards these majestic animals while reaping profitable benefits.

The presence or absence of hope is intricately linked to our environmental context, encompassing various facets, such as organisations, work, or home, in addition to the goals we set for ourselves, shaping our levels of hopefulness (Averill et al., 1990). Hope proves most effective when our goals hold personal significance, are aligned with social and moral standards, are perceived as attainable, and are within our control. When young individuals develop habitual pro-environmental behaviour due to parental influence, climate change may not be equally important. The alignment of values and goals nurtures hope towards goal achievement and is a powerful catalyst for personal growth and success in environmental stewardship.

The prevailing model of hope, conceptualising it as goal-directed thinking, has found widespread implementation and demonstrated efficacy in various contexts, such as the "Making Hope Happen" programme (Lopez, 2013) in schools. This research, however, goes beyond this model to explore the nuanced nature of hope. This initiative aims to educate students about hope, instil new beliefs and skills, and sustain their high motivation levels while designing plans to achieve environmental aspirations. By acknowledging the limitations of this model, we open up the possibility for a more comprehensive understanding of hope in the context of pro-environmental action.

Hope holds particular significance for young people within the context of pro-environmental action, determining positive youth development and correlating with life satisfaction (Marques et al., 2013), self-worth, improved school attendance, higher grades, physical health, and overall wellbeing (Snyder et al., 2018). Hopeful students eagerly anticipate their future, actively engage in environmental initiatives, and are less likely to neglect pro-environmental responsibilities.

In the face of adversity, practising hope emerges as one of individuals' most effective coping strategies (Lazarus, 1999). During challenging or unsatisfactory times in the pursuit of environmental goals, embracing hope allows us to envision a brighter future beyond the complexities of today.

Hope profiling (adapted from Lopez et al., 2004)

Write five stories about your past or current environmental goal pursuits. As you write your stories, please include information about how you developed your goals or paths you followed in working towards your goal and where your motivation to work on your goals came from. If you can't think of environmental pursuits, consider non-environmental ones. Feel free to write about goal pursuits in your various life domains.

Create 3 column on your paper. On the top of the first column write "Goals", on the top of the second column write "Pathways", on the top of the third column write "Obstacles". Under Goals write down a goal for using one of your strengths. In the Pathways column, write down at least three ways in which you can accomplish your goal or make using your strengths easier. In the obstacles' column, write down at least one obstacle for each pathway you mentioned.

Part 2

Reflect on how you will keep your motivation for accomplishing your environmental goal. Who can help you do it? What other resources can you activate to make it happen?

Here is an example of what your writing could look like:

Part 1: Goal Pursuits

1 **Goal:** Implementing a backyard composting system to reduce household waste.

- **Pathways**
  - i Research various composting methods online and select the most suitable one.
  - ii Attend a local workshop or seminar on composting to gain practical knowledge.
    - b Connect with experienced composters in online forums or community groups for guidance.
- **Obstacles**
  - i Limited space for setting up a composting system.
  - ii Lack of time to maintain the compost pile regularly.

2 **Goal:** Establishing a vegetable garden to promote biodiversity and reduce carbon footprint.

- **Pathways**
  - i Design a garden layout that considers sunlight exposure and soil quality.
  - ii Purchase organic seeds or seedlings from local nurseries or farmers' markets.
    - b Install a rainwater harvesting system to minimise water consumption.

- **Obstacles**
  - i Unpredictable weather conditions affect plant growth.
  - ii Pest infestations damage the crops.

3 **Goal:** Adopting eco-friendly transportation habits to reduce carbon emissions.

- **Pathways**
  - i Research public transportation routes and schedules for daily commuting.
  - ii Invest in a bicycle and explore bike-friendly routes for short-distance travel.
  - iii Organise a carpooling schedule with neighbours or co-workers for longer journeys.
- **Obstacles**
  - i Limited availability and reliability of public transportation options.
  - ii Safety concerns while cycling on busy roads.

Part 2: Maintaining Motivation
To sustain motivation for accomplishing these goals, it is essential to establish a support network and tap into available resources:

1 **Support System:** Engage family members, friends, or neighbours with similar environmental goals. Forming a support group or participating in community initiatives can provide encouragement and accountability.
2 **External Resources**

- **Online Communities:** Join forums or social media groups dedicated to sustainable living and environmental activism. These platforms offer a wealth of knowledge, inspiration, and practical advice.
- **Local Organisations:** Seek guidance from local environmental organisations or conservation groups. They may offer workshops, educational resources, and volunteer opportunities to support your efforts.
- **Government Programmes:** Explore government incentives or discounts for eco-friendly practices such as composting, gardening, or using public transportation. These programmes can provide financial assistance and additional motivation.

By leveraging support networks and external resources, individuals can stay motivated and overcome obstacles to achieve pro-environmental goals.

It is crucial to distinguish between optimism and hope in the context of pro-environmental action. While optimism entails the expectation that everything will work out well, hope provides "the will and the way" to persevere. In environmental action, optimism may lead to complacency, while hope spurs us to take active steps towards our goals. It comprises agency thinking and clear pathway thinking that collectively guide us. Agency thinking involves maintaining the energy to pursue our goals related to climate protection and sustaining motivation. It affirms that we can overcome obstacles and persist in our environmental endeavours. Pathway thinking involves developing a clear pathway towards achieving our environmental goals, outlining steps and planning actions to make them a reality. True hope goes beyond wishful thinking; it is a comprehensive plan and a belief that we possess the necessary resources to sustain the energy and ultimately achieve our environmental objectives.

In certain instances, presenting hopeful ideas in the context of pro-environmental actions may lead to passivity (Fielding & Hornsey, 2016); the interplay between hope and action is ultimately shaped by personal characteristics, message - framing, and the timeframe being considered. Therefore, when used effectively, hopeful messages envision a brighter future and provide a roadmap to achieve it.

Purpose and meaning-making

The concept of meaning in life encompasses various viewpoints. Some define it as realising personal goals (Ryff & Singer, 1998), while others view it as understanding the inherent order and purpose of existence (Reker & Wong, 1988). Leading a meaningful life often implies a sense of significance and coherence for individuals (Heintzelman & King, 2014). This prevailing definition emphasises maintaining internal consistency in self-understanding, having a clear life purpose (Steger et al., 2015), and comprehending how life experiences shape our current selves.

While meaning addresses the theoretical aspect of our experience, purpose involves its practical application. It is essential to distinguish between the meaning in life and the meaning "of" life, with the latter probing existential questions about our existence.

In the realm of pro-environmental, biodiversity, and climate action, understanding life's meaning is intertwined with a broader sense of purpose. Individuals who align their actions with environmental values, contribute to sustainability, and work towards a healthier planet often find profound meaning in their efforts. This involves recognising the significance of preserving biodiversity, mitigating climate change impacts, and promoting a sustainable coexistence with nature. The application of meaning in this context extends beyond personal fulfilment to embrace a collective purpose that addresses environmental challenges and enhances the wellbeing of future generations. The theoretical understanding of meaning aligns with practical engagement, highlighting the transformative potential of pro-environmental actions in shaping a purpose-driven and impactful life.

Exploring the meaning in life is crucial for our wellbeing, especially in the context of biodiversity, climate change, and pro-environmental action. Research indicates a strong link between understanding life's meaning and improved physical health, which protects against pathologies and contributes to overall wellbeing (Roepke et al., 2014; Steger, 2017). However, studies show mixed results concerning the relationship between pro-environmental behaviour and life meaning, depending on the conceptualisations used (e.g., Gu et al., 2015; Lutz et al., 2023). This highlights the need for further research, which could unveil new insights and deepen our understanding of these complex relationships.

The role of meaning in life significantly influences the connection between pro-environmental actions and overall wellbeing. Individuals actively engaged in such actions often report reduced loneliness, attributed to the profound meaning derived from their environmental involvement (Jia et al., 2021). Several key elements contribute to this connection. Firstly, pro-environmental individuals feel a profound responsibility to educate future generations about environmental issues, becoming a source of life's meaning and enhancing their overall wellbeing. A deep appreciation for nature significantly adds to the meaning of life experienced by those involved in pro-environmental actions. This connection is a source of inspiration, influencing their environmental engagement and alleviating feelings of loneliness.

Self-directed learning is pivotal in renewing agency among individuals engaged in pro-environmental actions. Taking the initiative to learn about environmental issues fosters empowerment and purpose, ultimately affecting their perception of loneliness. These findings highlight the vital role of meaning in life as a core motivator underlying the association between pro-environmental engagement and reduced loneliness. Embracing responsibilities for environmental education, fostering a deep connection to nature, and experiencing renewed agency through self-directed learning enable individuals to find profound meaning in their commitment to environmental actions, positively impacting their overall wellbeing and sense of social connectedness.

Environmental obituary (adapted from Frisch, 2006)

Visualise a scenario where your current unsustainable and environmentally harmful behaviours persist unchanged until the end of your life. Reflect on all the habits and routines contributing to environmental degradation, and consider the escalating consequences if these patterns worsen each year until your passing. Envision an extended timeline where you make no positive changes to your lifestyle. Your ecological standards, priorities, and goals remain stagnant, intensifying environmental degradation. Now, create a narrative envisioning your obituary under these circumstances. Make it personal and detailed, contemplating how your choices could impact the environment. Picture your family, friends, and strangers reading about your environmental legacy in newspapers and online. Describe the consequences of your actions on the natural world if you fail to alter your unsustainable habits.

Remarkably, the relationship between meaning and mood operates bidirectionally, as an improved emotional state has been demonstrated to enhance our sense of meaning in life (Hicks & King, 2009). This interplay becomes especially vital during challenging times, where life meaning acts as a guiding force, offering a unique form of mental health protection that surpasses the benefits derived from positive emotions alone (Wong, 2011). Understanding life's meaning becomes instrumental in helping individuals navigate and find purpose amid environmental adversities, contributing to mental resilience.

It is crucial to note that while having a sense of life meaning is associated with higher levels of wellbeing, actively searching for meaning can have an inverse effect (Li et al., 2020). This insight highlights the importance of a balanced and reflective approach to understanding one's purpose, particularly in the ever-evolving landscape of environmental concerns.

Life purpose takes on added significance in biodiversity, climate change, and pro-environmental action (Steger, 2012) as a motivational and practical dimension of meaning. For example, an individual whose meaning in life revolves around personal growth may identify becoming an advocate for environmental sustainability as a means to actualise this purpose. Engaging in continuous learning and growth alongside collective efforts for biodiversity conservation aligns with this individual's life purpose, providing a sense of direction and fulfilment.

Life purpose becomes intricately linked to the goals we set for ourselves, making it a driving force behind pro-environmental action. Without purpose, even if individuals possess a clear sense of meaning, they might lack the strength or courage to pursue environmentally conscious behaviours actively. On the contrary, having a life purpose without a clear understanding of meaning may lead to a realisation later in life that actions and values are incongruent with personal convictions. The optimal scenario is to simultaneously understand our meaning and purpose, aligning our actions with a more profound sense of significance and direction in biodiversity, climate change, and pro-environmental endeavours. This integrated approach fosters individual wellbeing and contributes to a collective sense of purpose in addressing environmental challenges.

Logotherapy

During the World War II, Victor Frankl (2000), a psychiatrist who found himself captured by the Nazis and confined to a concentration camp, embarked on a profound exploration of human resilience. Amidst the stark realities of imprisonment, he observed the contrasting reactions of fellow prisoners. While some succumbed to early despair, others clung to hope and resilience, resisting their oppressors. In the post-war era, some individuals reported flourishing while others grappled with the challenges of everyday life. Frankl identified a pivotal factor that set these groups apart: the presence of life meaning.

This revelation fuelled the development of Logotherapy, a therapeutic approach that Frankl created. Unlike conventional therapeutic methods that primarily target wellbeing or depression, Logotherapy delves into the profound journey of exploring and uncovering life meaning. Frankl's findings showed the transformation that emerged when individuals, particularly former prisoners of war, gained awareness of the profound meaning of their suffering. The realisation of this meaning became a catalyst for a significant shift in their lives. The paths to discovering this meaning varied among individuals – some traversing it over the years, while others experienced this profound transformation in a single therapeutic session.

Logotherapy and Victor Frankl explored the relationship between meaning in life, pro-environmental intentions, and behaviours. The findings suggest that meaning in life plays a crucial role in driving pro-environmental behaviours. This implies that individuals with a strong sense of purpose and meaning are more likely to engage in actions that contribute to environmental sustainability.

Understanding the importance of meaning in life as a motivator for pro-environmental behaviours has several implications. Firstly, it highlights the need to incorporate discussions about meaning and purpose into environmental education and outreach programmes. By helping individuals connect their values and aspirations to environmental stewardship, educators and advocates can enhance engagement and commitment to pro-environmental actions.

Secondly, organisations and policymakers involved in environmental initiatives should recognise the significance of meaning in life as a driving force behind behaviour change. Strategies and interventions promoting sustainable behaviours may be more effective if they tap into individuals' intrinsic motivations and a sense of purpose.

Moreover, the study highlighted the interconnectedness of psychological factors, such as meaning in life, with environmental attitudes and behaviours. This suggests that addressing individuals' psychological wellbeing and sense of fulfilment may positively influence their environmental actions.

Overall, the study's implications emphasise the importance of considering meaning in life as a fundamental aspect of motivation for pro-environmental behaviours. By acknowledging and harnessing the power of meaning, stakeholders in environmental conservation efforts can potentially enhance their effectiveness in promoting sustainable behaviours and fostering a deeper connection between individuals and the natural world.

Positive environmental legacy (adapted from Rashid & Seligman, 2019)

Shift your focus towards the future and reflect on how you aspire to be remembered by your family, friends, colleagues, and everyone close to you in the context of your environmental impact. What environmental legacy do you want to leave behind? What do you hope people will say about your contributions to the planet? Dedicate 15–20 minutes to articulate your aspirations. Once completed, review your vision and formulate a plan to bring this positive environmental legacy to fruition. Identify the changes needed in your current lifestyle and set tangible goals to guide your journey. Regularly revisit your vision to stay motivated and on track, ensuring that your actions align with the positive environmental legacy you envision.

Likewise, engaging in pro-environmental actions can serve as a powerful catalyst for discovering meaning in life. Conversely, logotherapy can be employed to assist individuals in finding purpose in their pro-environmental endeavours, thereby enhancing their overall wellbeing. Consider the following examples illustrating how this can be achieved.

For instance, individuals can reflect on their inner values and sources of meaning, tying personal values to nature, sustainability, and the wellbeing of future generations. Connecting eco-friendly behaviours to a broader purpose, such as contributing to a healthier planet or preserving biodiversity, allows individuals to find deeper meaning in their environmental actions.

Logotherapy emphasises taking responsibility for one's choices, aligning with individuals recognising their role as stewards of the Earth. Acknowledging the impact of current actions on future generations can further motivate individuals to make choices that positively influence the environment.

Turning environmental challenges into opportunities involves viewing adversity as a chance for personal growth, learning, and contributing to positive change. Engaging in sustainable practices can become a path to personal development, fostering a sense of accomplishment and purpose.

Logotherapy recognises the significance of finding meaning in various aspects of life, including nature. It encourages individuals to establish a meaningful connection with the natural world. Incorporating nature-based activities into logotherapeutic interventions, such as ecotherapy or outdoor mindfulness practices, enhances individuals' appreciation for nature and strengthens their resolve to protect it.

Embracing voluntary simplicity, a lifestyle prioritising meaningful experiences over material possessions, aligns with logotherapy's emphasis on intentional living. Making intentional choices to reduce environmental impact can further promote a sense of purpose derived from sustainable practices.

Highlighting the importance of community and shared values, logotherapy encourages communities to embrace environmental values collectively, fostering a shared sense of meaning and purpose. Creating opportunities for collective pro-environmental activities within communities can enhance individuals' sense of belonging and purpose.

Educational interventions involve enlightening individuals about the potential for meaning in various aspects of life. Applying this to environmental concerns requires educational programmes that highlight the meaningful impact of pro-environmental actions, they can inspire informed choices, and reinforce a sense of purpose through conscious decision-making.

Incorporating logotherapeutic principles into ecotherapy sessions provides individuals a structured approach to finding meaning in their relationship with the environment. Combining logotherapy with mindfulness practices in natural settings can enhance individuals' connection to the environment, promoting a sense of purpose through mindful engagement with nature.

Ultimately, the principles of logotherapy offer a comprehensive framework for individuals to derive meaning and purpose from their pro-environmental actions, contributing to a sustainable and ecologically conscious way of life.

Conclusion

This chapter has highlighted the growing body of evidence indicating a significant impact of overall pro-environmental behaviour on psychological wellbeing and vice versa. However, it has also indicated a nuanced understanding of the relationship between pro-environmental activism and psychological wellbeing concerning various psychological wellbeing elements. Additionally, the chapter has shed light on the emerging evidence linking flow, a crucial component of psychological wellbeing, with environmental actions while also acknowledging the current gap in research in this area.

Moreover, the chapter has explored the concepts of accomplishment, anticipation, hope, and optimism as significant contributors to psychological wellbeing, with correlations between these factors and pro-environmental action. This suggests promising avenues for future research and intervention strategies to enhance environmental engagement and psychological wellbeing.

Finally, the discussion has delved into the roles of purpose and meaning, highlighting their importance in shaping pro-environmental intentions and activism and their potential to improve psychological wellbeing. This emphasised the interconnectedness between environmental values, personal fulfilment, and mental health outcomes.

The next chapter will delve into mindsets that can either hinder or foster emotional and psychological wellbeing, providing further insights into the intricate relationship between environmental actions and mental health.

References

Averill, J. R., Catlin, G., & Chon, K. K. (1990). *Rules of hope*. Springer-Verlag Publishing. https://doi.org/10.1007/978-1-4613-9674-1

Bandura, A. (1977). Self-efficacy: Toward a unifying theory of behavioral change. *Psychological Review*, 84(2), 191–215. https://doi.org/10.1037/0033-295X.84.2.191

Blickley, J. L., et al. (2013). Graduate student's guide to necessary skills for nonacademic conservation careers. *Conservation Biology: The Journal of the Society for Conservation Biology*, 27(1), 24–34. https://doi.org/10.1111/j.1523-1739.2012.01956.x

Bryant, S. E. (2003). The role of transformational and transactional leadership in creating, sharing and exploiting organizational knowledge. *Journal of Leadership & Organizational Studies*, 9(4), 32–44. https://doi.org/10.1177/107179190300900403

Carr, A. (2020). *Positive psychology and you: A self-development guide*. Routledge.

Chowdhury, R., Sharot, T., Wolfe, T., Düzel, E., & Dolan, R. J. (2014). Optimistic update bias increases in older age. *Psychological Medicine*, 44(9), 2003–2012. Doi: 10.1017/S0033291713002602

Corral-Verdugo, V., Montiel-Carbajal, M. M., Sotomayor-Petterson, M., Frías-Armenta, M., Tapia-Fonllem, C., & Fraijo-Sing, B. (2013). Psychological wellbeing as correlate of sustainable behaviors. In C. García, V. Corral-Verdugo, & D. Moreno (Eds.), *Recent Hispanic research on sustainable behavior and interbehavioral psychology* (pp. 27–40). Nova Science Publishers.

Csikszentmihalyi, M. (1997). *Creativity: Flow and the psychology of discovery and invention*. HarperCollins Publishers.

Evans, L., Maio, G. R., Corner, A., Hodgetts, C. J., Ahmed, S., & Hahn, U. (2017). Self-interest and pro-environmental behaviour. *Nature Climate Change*, 3(2), 122–125.

Fahlman, S. A., Mercer-Lynn, K. B., Flora, D. B., & Eastwood, J. D. (2013). Development and Validation of the Multidimensional State Boredom Scale. *Assessment*, 20, 68–85.

Fielding, K. S., & Hornsey, M. J. (2016). A social identity analysis of climate change and environmental attitudes and behaviors: Insights and opportunities. *Frontiers in Psychology*, 7, Article 121. https://doi.org/10.3389/fpsyg.2016.00121

Frankl, V. (2000). *Man's search for ultimate meaning*. MJF Books.

Frisch, M. B. (2006). *Quality of life therapy: Applying a life satisfaction approach to positive psychology and cognitive therapy*. John Wiley & Sons Ltd.

Fullagar, C. J., & Fave, A. D. (2017). *Flow at work: Measurement and implications, Current issues in work and organizational psychology*. Routledge/Taylor & Francis Group.

Gilbert, D. T., & Wilson, T. D. (2007). Prospection: Experiencing the future. *Science*, 317(5843), 1351–1354. https://doi.org/10.1126/science.1144161

Gu, D., Huang, N., Zhang, M., & Wang, F. (2015). Under the dome: Air pollution, wellbeing, and pro-environmental behaviour among beijing residents. *Journal of Pacific Rim Psychology*, 9(2), 65–77. https://doi.org/10.1017/prp.2015.10

Han, G.-S. (2023). Relationships between outdoor recreation-associated flow, pro-environmental attitude, and pro-environmental behavioral intention. *Sustainability*, 15(13), 10581. https://doi.org/10.3390/su151310581

Hefferon, K., & Boniwell, I. (2011). *Positive psychology: Theory, research and applications*. McGraw-Hill.

Heintzelman, S. J. S. M. M. E., & King, L. A. (2014). Life is pretty meaningful. *American Psychologist*, 69, 561–574.

Hicks, J. A., & King, L. A. (2009). Positive mood and social relatedness as information about meaning in life. *Journal of Positive Psychology*, 4, 471–482.

Huppert, F. A. (2009). Psychological well-being: Evidence regarding its causes and consequences. *Applied Psychology: Health and Well-Being*, 1(2), 137–164. https://doi.org/10.1111/j.1758-0854.2009.01008.x

Ilies, R., Wagner, D., Wilson, K., Ceja, L., Johnson, M., DeRue, S., & Ilgen, D. (2017). Flow at work and basic psychological needs: Effects on well-being. *Applied Psychology: An International Review*, 66(1), 3–24. https://doi.org/10.1111/apps.12075

Jia, F., Soucie, K., Matsuba, K., & Pratt, M. W. (2021). Meaning in life mediates the association between environmental engagement and loneliness. *International Journal of Environmental Research and Public Health*, 18(6), 2897. https://doi.org/10.3390/ijerph18062897

Lazarus, R. S. (1999). Hope: An emotion and a vital coping resource against despair. *Social Research*, 66, 653–678.

Li, J.-B., Dou, K., & Liang, Y. (2020). The relationship between presence of meaning, search for meaning, and subjective well-being: A three-level meta-analysis based on the meaning in life questionnaire. *Journal of Happiness Studies: An Interdisciplinary Forum on Subjective Well-Being,* 22(1), 467–489. https://doi.org/10.1007/s10902-020-00230-y

Liu, Y., Cleary, A., Fielding, K. S., Murray, Z., & Roiko, A. (2022). Nature connection, pro-environmental behaviours and wellbeing: Understanding the mediating role of nature contact. *Landscape and Urban Planning*, 228, 104550. https://doi.org/10.1016/j.landurbplan.2022.104550

Locke, E. A., & Latham, G. P. (2002). Building a practically useful theory of goal setting and task motivation: A 35-year odyssey. *American Psychologist*, 57(9), 705–717.

Lopez, S. J. (2013). *Making hope happen: Create the future you want for yourself and others*. Atria Books.

Lopez, S. J., et al. (2004). Strategies for accentuating hope. In P. A. Linley & S. Joseph (Eds.), *Positive psychology in practice* (pp. 388–404). John Wiley & Sons, Inc. https://doi.org/10.1002/9780470939338.ch24

Lutz, P. K., Zelenski, J. M., & Newman, D. B. (2023). Eco-anxiety in daily life: Relationships with well-being and pro-environmental behavior. *Current Research in Ecological and Social Psychology*, 4. https://doi.org/10.1016/j.cresp.2023.100110

MacKinnon, M., Davis, A. C., & Arnocky, S. (2022). Optimistic environmental messaging increases state optimism and in vivo pro-environmental behavior. *Frontiers in Psychology*, 13, 856063. https://doi.org/10.3389/fpsyg.2022.856063

Marques, S., Lopez, S., & Mitchell, J. (2013). The role of hope, spirituality and religious practice in adolescents' life satisfaction: Longitudinal findings. *Journal of Happiness Studies*, 14, 251–261.

Melges, F. T. (1990). Identity and temporal perspective. In R. A. Block (Ed.), *Cognitive models of psychological time* (pp. 255–266). Lawrence Erlbaum Associates, Inc.

Rashid, T., & Seligman, M. (2019). *Positive psychotherapy workbook*. Oxford University Press.

Reker, G. T., & Wong, P. T. P. (1988). Aging as an individual process: Toward a theory of personal meaning. In J. E. Birren & V. L. Bengtson (Eds.), *Emergent theories of aging* (pp. 214–246). Springer Publishing Co.

Roepke, A. M., Benson, L., Tsukayama, E., & Yaden, D. B. (2018). Prospective writing: Randomized controlled trial of an intervention for facilitating growth after adversity. *The Journal of Positive Psychology*, 13(6), 627–642.

Roepke, A. M., Jayawickreme, E., & Riffle, O. M. (2014). Meaning and health: A systematic review. *Applied Research in Quality of Life*, 9, 1055–1079.

Ryff, C. D. (1989). Happiness is everything, or is it? Explorations on the meaning of psychological well-being. *Journal of Personality & Social Psychology*, 57, 1069–1081.

Ryff, C. D., & Singer, B. (1998). The contours of positive human health. *Psychological Inquiry*, 9, 1.

Seligman, M. E. P. (1998). *Learned optimism: How to change your mind and your life*. Free Press.

Seligman, M. E., Railton, P., Baumeister, R. F., & Sripada, C. (2016). *Homo prospectus*. Oxford University Press.

Sharot, T., Korn, C. W., & Dolan, R. J. (2011). How unrealistic optimism is maintained in the face of reality. *Nature Neuroscience*, 14(11), 1475–1479. https://doi.org/10.1038/nn.2949

Sharot, T., Riccardi, A. M., Raio, C. M., & Phelps, E. A. (2007). Neural mechanisms mediating optimism bias. *Nature*, 450(7166), 102–105. https://doi.org/10.1038/nature06280

Sheldon, K. M., & Elliot, A. J. (1999). Goal striving, need satisfaction, and longitudinal well-being: The self-concordance model. *Journal of Personality and Social Psychology*, 76(3), 482–497.

Snyder, C. R., Rand, K. L., & Sigmon, D. R. (2018). Hope theory: A member of the positive psychology family. In M. W. Gallagher & S. J. Lopez (Eds.), *The Oxford handbook of hope* (pp. 257–276). Oxford University Press.

Steger, M., Fitch-Martin, A., Donnelly, J., & Rickard, K. (2015). Meaning in life and health: Proactive health orientation links meaning in life to health variables among american undergraduates. *Journal of Happiness Studies*, 16, 583–597.

Steger, M. F. (2012). Experiencing meaning in life: Optimal functioning at the nexus of well-being, psychopathology, and spirituality. In P. T. P. Wong (Ed.), *The human quest for meaning: Theories, research, and applications* (2nd ed., pp. 165–184). Routledge/Taylor & Francis Group.

Steger, M. F. (2017). Meaning in life and wellbeing. In M. Slade, L. Oades, & A. Jarden (Eds.), *Wellbeing, recovery and mental health* (pp. 75–85). Cambridge University Press.

Tímea, M., & Attila, O. (2020). A társas helyzetben tapasztalt flow-élmény kapcsolata az élettel való elégedettséggel és a pszichológiai jólléttel, fiatal felnőtteknél [Experiencing flow in social activities and its relationship with satisfaction with life and psychological wellbeing, in the sample of young adults]. *Mentálhigiéné és Pszichoszomatika*, 21(1), 37–55. https://doi.org/10.1556/0406.21.2020.003

Tse, D. C. K., Lau, V. W.-Y., Perlman, R., & Mclaughlin, M. (2020). The development and validation of the Autotelic Personality Questionnaire. *Journal of Personality Assessment*, 102, 88–101.

Wong, P. T. P. (2011). Positive psychology 20: Towards a balanced interactive model of the good life. *Canadian Psychology/Psychologie Canadienne*, 52, 69–81.

7 Mindsets as guideposts for wellbeing

In the city's bustling heart, Sarah carefully scrutinises the labels of beauty products lining the shelves of her local grocery store. With an environmentally conscious mindset guiding her choices, she seeks out items with minimal packaging, opting for those made from sustainable materials. As she fills her shopping basket, her thoughts drift to the nearby farms that supply the store's produce. Choosing locally sourced fruits and vegetables reduces her carbon footprint and supports the hardworking farmers in her community.

On the other side of town, a group of like-minded individuals converge at the edge of a forest, armed with gloves and garbage bags. They share an understanding of the ecosystems and the need to preserve it. Together, they embark on a community clean-up initiative, each discarded plastic bottle or stray piece of litter they collect symbolising a step towards preserving the delicate balance of local flora and fauna. For them, biodiversity isn't just a concept – it's a cause worth fighting for, one clean-up at a time.

Marcus is a vocal advocate for environmental reform. His proactive environmental mindset drives him to tirelessly lobby policymakers for stronger regulations on emissions and wildlife protection. He recognises that real change necessitates systemic action, and he's resolute in his mission to hold those in positions of influence accountable for safeguarding the planet for future generations.

Meanwhile, a dedicated group of educators and activists are hard at work in classrooms and community centres. Through workshops and educational campaigns, they seek to ignite a spark of environmental awareness in the hearts and minds of those around them. Their mission is clear: to inspire others to embrace pro-environmental behaviours and cultivate a deeper understanding of biodiversity conservation.

In the quiet moments of everyday life, Tom and Emily make small but meaningful changes. Embracing a sustainable lifestyle mindset, they opt for public transportation over cars and invest in renewable energy

DOI: 10.4324/9781003452676-9

sources like solar panels. With each conscious choice, they're actively minimising their ecological footprint and paving the way towards a more sustainable future.

From the aisles of the grocery store to the halls of power, from community clean-ups to educational workshops, these individuals are united by a shared commitment to protecting the planet. With their diverse actions and mindsets, they're each playing a vital role in the collective effort to safeguard our environment for generations to come.

In the vast domain of environmental stewardship, a diverse spectrum of mindsets exists, each carrying its weight of influence on individual and collective pro-environmental behaviour. However, within the scope of this chapter, our attention pivots towards those specific mindsets intricately intertwined with wellbeing – the attitudes and perspectives that serve as either catalysts or barriers to personal fulfilment and contentment. These mindsets impact individual wellbeing and influence the level of engagement in actions aimed at benefiting the environment.

As we navigate these mindsets, we embark on a journey to unravel their subtleties, understand their far-reaching impacts, and uncover their latent potential for transforming wellbeing and sustainability efforts. By delving into the intricate interplay between these mindsets and pro-environmental conduct, we aim to explore pathways leading towards a future characterised by sustainability and harmonious coexistence between humanity and the natural world. Understanding these mindsets can create positive change in our relationship with the environment.

We will begin this chapter by delving into the passion for environmental activism, often regarded as a driving force for environmental enthusiasts. However, it is a double-edged sword, capable of unexpectedly impacting our wellbeing negatively. Next, we explore the concept of a growth mindset and its potential to amplify pro-environmental actions. Following this, we delve into stress mindset research and its relevance in aiding individuals experiencing eco-anxiety and stress linked to climate change.

We further investigate the mindsets associated with our time perspectives, which can either impede or facilitate our engagement in pro-environmental action, affecting our wellbeing positively or negatively. Finally, we delve into maximising and satisfying thinking, which significantly influences our happiness and contentment with our actions to protect the planet. By understanding these diverse mindsets, we can gain valuable insights into how they shape our overall wellbeing and, our capacity to contribute positively to environmental conservation endeavours.

Passion

Numerous books explore the virtues of passion and its pivotal role in enhancing our lives and performance. Consequently, there's a prevailing assumption that passion is unequivocally beneficial. However, research suggests that passion can have positive and negative implications (Vallerand, 2003). In some cases, individuals with passion for a particular pursuit may be less inclined to invest additional effort in honing their skills and deepening their expertise (O'Keefe et al., 2018).

For instance, individuals who passionately advocate for conservation efforts may become complacent, believing their passion alone can drive change. As a result, they may be less inclined to delve deeper into understanding the intricacies of environmental issues or to actively seek out innovative solutions. This can hinder progress towards achieving meaningful and lasting success in their conservation endeavours. Furthermore, the belief that passion should inherently sustain interest may lead some environmental activists to abandon causes when faced with obstacles or challenges instead of persevering and adapting their approaches. These are just a few adverse outcomes of passion.

According to the Duality of Passion Model (Vallerand, 2003), passion can be categorised into two primary types: harmonious and obsessive. Despite both types aiming for similar outcomes, their impact on wellbeing differs significantly. Let's examine these passion types within the context of environmental stewardship.

Harmonious passion for environmental stewardship entails a deep and intrinsic commitment to conservation efforts. Individuals with harmonious passion experience joy and fulfilment from their engagement in pro-environmental actions. Their dedication to environmental causes is integrated seamlessly into their identity, enriching their lives and contributing positively to their wellbeing. For example, a person with harmonious passion may find immense satisfaction in volunteering for habitat restoration projects or participating in community clean-up initiatives. Their passion for environmental stewardship enhances their sense of purpose and fulfilment, fostering a sustainable and fulfilling lifestyle.

On the other hand, an obsessive passion for environmental stewardship manifests as a rigid and uncontrollable urge to engage in conservation activities. Individuals with obsessive passion may experience feelings of pressure and guilt if they cannot devote sufficient time and energy to environmental causes. Their passion for environmental stewardship may become all-consuming and detrimental to their mental and emotional wellbeing. For instance, someone with obsessive passion may experience burnout from overcommitting to environmental activism or struggle with feelings of inadequacy if they perceive their efforts as falling short of their expectations. In extreme cases, obsessive passion may lead to detrimental behaviours, such as neglecting personal relationships or neglecting self-care in favour of environmental activism.

While both harmonious and obsessive passion for environmental stewardship aims to achieve similar outcomes, their differing impacts on wellbeing highlight the importance of cultivating a healthy and balanced approach to environmental activism. By fostering harmonious passion and integrating environmental stewardship into a fulfilling and sustainable lifestyle, individuals can contribute meaningfully to conservation efforts while prioritising their wellbeing.

Here are some of the defining features of both types of passion:

Harmonious Passion involves a robust and flexible inclination towards an activity that individuals genuinely love and willingly choose to engage in, with the harmonious integration of the activity into one's identity.

Key Characteristics

- **Enjoyment:** Individuals with a harmonious passion for environmental stewardship derive genuine enjoyment and fulfilment from their engagement in conservation efforts. They find intrinsic value in habitat restoration, sustainable living practices, and community clean-up initiatives.
- **Autonomy:** Participation in pro-environmental actions is voluntary and aligned with personal values. Individuals with harmonious passion actively engage in environmental stewardship, recognising its significance and contribution to wellbeing.
- **Flexibility:** Despite their strong commitment to environmental causes, individuals with harmonious passion can maintain a healthy balance in their lives. They can detach from environmental activism when necessary without experiencing internal conflict or negative emotions. This flexibility allows them to prioritise self-care and maintain sustainable engagement in pro-environmental activities over the long term.
- **Positive Outcomes:** Harmonious passion for pro-environmental stewardship is associated with positive outcomes, including greater life satisfaction, overall wellbeing, and positive affect. Individuals who embody harmonious passion experience a sense of purpose and fulfilment from their contributions to environmental conservation efforts, enriching their lives and enhancing their sense of connectedness to nature and community.

Obsessive Passion entails a rigid and uncontrollable urge to engage in an activity, characterised by relentless internal pressure to engage, often at the expense of other life domains.

Key Characteristics

- **Compulsion:** Individuals with obsessive passion experience an intense internal pressure or obsession to participate in environmental activism. This compulsion may lead to a sense of loss of control, as they feel compelled to prioritise environmental causes above all else.
- **Conflict:** Engagement in pro-environmental actions may conflict with other aspects of individuals' lives, such as work, family, or personal

interests. This conflict can result in adverse outcomes, including heightened stress levels, feelings of guilt, and difficulty maintaining a healthy work-life balance.

- **External Rewards:** Individuals driven by obsessive passion may be motivated primarily by external rewards, such as social approval or validation from peers or society at large. External validation fuels their focus on environmental activism rather than an intrinsic passion for conservation efforts.
- **Opposing Outcomes:** Obsessive passion for pro-environmental stewardship is associated with adverse consequences, including burnout, heightened anxiety, and lower overall levels of wellbeing. Individuals may experience emotional exhaustion and fatigue from their relentless pursuit of environmental activism, ultimately compromising their long-term sustainability of meaningful engagement in conservation efforts.

Let us illustrate this with an example. Grace is a 20-something environmental activist. Over the years, Grace's transformation through her unwavering commitment to environmental causes has been remarkable. She not only adjusted her behaviour to adhere to sustainability principles but also passionately persuaded her family and friends to join the cause. Recognising the pivotal role of widespread awareness in sparking transformative change, Grace decided to amplify her efforts. Venturing beyond conventional strategies, she became part of a group of innovative activists, engaging in unconventional actions like hunger strikes and picketing outside the parliament. Immersed in activities that fuelled her passion, Grace sought to make a lasting impact on the environmental front, a testament to the power of individual action.

As Grace delved deeper into her activism, she found profound meaning in her life. However, a few months into her passionate endeavours, she encountered challenges many of us can relate to. She found it difficult to fall asleep and maintain her sleep and the joy she once derived from life seemed to fade away. Grace initially thought that her poorer wellbeing was due to the lack of action of the governing bodies, but soon she realised that her enthusiasm for activism was causing unease rather than the expected fulfilment. Despite living the life she aspired to lead, she became deeply unhappy in our pursuits.

Gradually, the Duality of Passion model is entering the research on pro-environmental behaviour. For instance, a study involving academics promoting a green work climate revealed that a harmonious passion for the environment was a stronger predictor of pro-environmental behaviour than having an obsessive passion (Choong et al., 2020). Another study examined outdoor activities and their impact on pro-environmental actions (Junot et al., 2017). It found that harmonious passion was associated with positive emotions, which, in turn, were linked to a sense of connection with nature and, ultimately, more frequent engagement in environmental behaviours. In contrast, obsessive passion was associated with negative emotions, which were inversely related to the sense of connection with nature and pro-environmental behaviours.

As we navigate the complexities of pro-environmental behaviour, we must reflect on the nature of our passion and take proactive measures to safeguard our wellbeing. This entails recognising whether our passion for environmental causes aligns more with the harmonious or obsessive spectrum and adjusting our approach accordingly. Furthermore, it serves as a call to action for leaders within the environmental movement to promote a balanced approach among their members. They must ensure that their initiatives and actions foster harmonious passion while mitigating the risk of promoting obsessive tendencies. This can be achieved by providing feedback that emphasises harmonious passion's positive facets while discouraging obsessive passion behaviours. By fostering a culture of balance and self-awareness, we can create a sustainable environment where passion for environmental stewardship thrives without sacrificing the wellbeing of those who champion the cause.

For over two decades, expressive writing has stood out as a transformative practice, offering numerous benefits for psychological wellbeing and can become a powerful tool for creating awareness of passion for environmental causes. This simple yet profound activity has been consistently linked to significant improvements in mental health (Pennebaker & Evans, 2014). Moreover, research suggests that expressive writing can effectively alleviate anxiety (Sloan et al., 2007) and reduce depressive symptoms (Pennebaker & Chung, 2011). A recent scoping review of interventions for eco-anxiety identified creative writing as a component for alleviating such concerns (Baudon & Jachens, 2021).

The positive impacts of expressive writing extend beyond emotional wellbeing, also influencing mood dynamics (Pennebaker, 1997) and even impacting physical health, such as improvements in immune functioning (Pennebaker et al., 1988), emphasising its role in promoting both health and reducing ill-being.

Environmental expressive writing (adapted from Pennebaker, 1997)

Grab a pen and a piece of paper. For the next 20 minutes, write about your deepest thoughts and feelings regarding your passion for environment. Dive into your innermost emotions and reflections connected to your commitment to environmental conservation. Repeat this activity daily for the next 3–5 days.

When employed to articulate our passion for climate action, expressive writing can offer valuable insights into the nature of our dedication. This understanding can then inform proactive steps to safeguard our wellbeing. The practice cultivates self-awareness and empowers individuals to protect their mental and physical health while actively advocating for environmental conservation.

Expressive writing's profound benefit in ecological wellbeing lies in its ability to tap into our subconscious, allowing our deepest thoughts and emotions regarding biodiversity and pro-environmental actions to surface. In a compelling U.S. experiment by Spera et al. (1994), participants facing job layoffs engaged in expressive writing interventions. Despite the study's shortened duration, over 50% of those participating in expressive writing discovered new opportunities, contrasting starkly with the non-writing group's outcome. Notably, both groups attended similar job-related events, highlighting the importance of expressive writing in providing an outlet for processing pent-up negative emotions and dealing with grief associated with significant life changes. This type of grief may resonate with individuals passionate about the environment, grappling with feelings of loss and mourning over climate change and environmental devastation.

Environmental stream of consciousness writing (adapted from Pennebaker & Evans, 2014)

This exercise is beneficial for individuals struggling with writing. Start writing using your a journal, or loose-leaf paper. Write stream-of-consciousness for ten minutes or until approximately two pages are filled. During your writing, track your thoughts and feelings as they arise. Write about your thoughts, feelings, hearing, smelling, or noticing. It's essential that your writing follows your stream of thought without worrying about spelling, grammar, or sentence structure. Remember, this writing is solely for your benefit, and you can discard it when finished. Just begin writing and keep going without pausing. Here are some prompts to help you start:

- I'm finding it challenging to write about the environment and my passion for it. What's causing this difficulty?
- Reflecting on this topic brings up a variety of disconnected thoughts, including...
- There have been instances in my life when I've experienced writer's block. How does this experience compare to those times? What's preventing me from getting started?
- This topic evokes numerous emotions in me. Some of these emotions include...

Growth mindset

Maya lived in a city surrounded by concrete buildings and polluted air. Despite the urban chaos, Maya had always felt a deep connection to nature and harboured a passion for environmental conservation.

Maya's journey was not without its challenges. Her efforts to make a difference were often met with frustration and disappointment. She volunteered for local clean-up events and advocated for sustainable practices in her community, but it seemed like no matter how hard she tried, the environmental problems only worsened. This uphill battle was a constant source of discouragement for Maya.

Over time, Maya began to develop a fixed mindset. She believed that her actions would make a slight difference and that the environmental challenges were insurmountable. She felt disheartened and discouraged, questioning whether her efforts were worth it.

One day, Maya attended a seminar on cultivating a growth mindset in environmental stewardship. The speaker shared stories of individuals who had transformed their communities through perseverance and resilience despite facing impossible odds.

Inspired by these stories, Maya began reassessing her approach to environmental activism. She realised that her fixed mindset was holding her back from realising her full potential and making a meaningful impact. She decided to adopt a growth mindset instead.

With her newfound mindset, Maya approached environmental challenges with renewed determination and optimism. She embraced setbacks as opportunities for learning and growth, viewing them as stepping stones rather than roadblocks. Maya collaborated with like-minded individuals and sought innovative solutions to environmental issues, refusing to be deterred by setbacks or obstacles.

As Maya continued to practise a growth mindset, she saw a remarkable transformation in her community. Her efforts to promote sustainability gained momentum, inspiring others to join her cause. Together, they implemented initiatives to reduce waste, conserve energy, and protect local ecosystems.

Through her unwavering commitment and belief in the power of change, Maya witnessed a significant positive impact on the environment and her community. Her journey from a fixed mindset to a growth mindset was a transformative one, not only in her approach to environmental stewardship but also in inspiring others to believe in their ability to create a better, more sustainable world.

The concept of a growth mindset (Dweck, 1998) has revolutionised how individuals motivate themselves to pursue improvement and persevere in the face of challenges. In the realm of environmental stewardship, where we often encounter discouraging statistics and obstacles, maintaining a growth mindset is not just beneficial; it's crucial. It enables us to remain resilient, dust ourselves off, and continue our efforts towards positive change. With a growth mindset, we embrace challenges as opportunities for growth, view setbacks as temporary setbacks, and persistently strive to make a difference in environmental conservation, despite the odds.

Embracing a growth mindset means believing in the potential for personal development and intellectual growth through perseverance and continuous learning. Challenges are seen as opportunities for growth, and effort is recognised as the path to mastery. People with a growth mindset welcome setbacks as opportunities to learn and improve and value constructive criticism as a means of self-improvement. They draw inspiration from the successes of others as motivation to keep pushing forward.

On the other hand, a fixed mindset involves believing that one's abilities and intellect are fixed and unchangeable. Individuals with a fixed mindset may avoid challenges to protect their sense of competence, viewing effort as futile. Setbacks are met with despair, and constructive criticism is often dismissed. Fear of failure looms, leading those with a fixed mindset to avoid situations challenging their perceived abilities.

Dweck's research emphasises that these mindsets are not inherent traits but rather attitudes that can be changed over time. Embracing a growth mindset opens up a journey of discovery, fostering resilience and adaptability in navigating life's ups and downs. In contrast, a fixed mindset may hinder personal and academic growth by instilling fear and reluctance to step outside one's comfort zone.

Extending Dweck's mindset theory to the complex realms of climate change, biodiversity, sustainability, and ecological stewardship provides valuable insights into how attitudes shape responses to environmental challenges. Within this intricate framework, the distinction between growth and fixed mindsets reveals clear patterns:

Those embracing a growth mindset see environmental challenges as opportunities for personal growth and evolution. They view climate change as a catalyst for self-discovery and take on the role of stewards, adopting sustainable practices to contribute meaningfully to environmental preservation. The resilience and determination are inherent in a growth mindset drive them forward despite the immense challenges of sustainability and climate action.

In contrast, individuals trapped in a fixed mindset may perceive climate change as an insurmountable and unchangeable aspect of the human condition. This mindset fosters helplessness and indifference, leading to apathy and resignation in the face of environmental imperatives. Pro-environmental actions are seen as futile, with individuals believing their efforts are insignificant against ecological challenges.

A growth mindset offers a valuable perspective on environmental education and awareness as catalysts for personal growth. Those with this mindset seek knowledge about ecological issues and continuously explore innovative solutions to integrate sustainability into daily life. Their thirst for understanding drives them to contribute to environmental conservation and coexistence with nature. Conversely, individuals in a fixed mindset resist change and cling to outdated habits, believing their ecological footprint is predetermined. They may disregard feedback about the environmental impact of their actions, viewing it as a threat to their identity.

Applying Dweck's mindset theory to climate change and pro-environmental action highlights a significant truth: a growth mindset fosters proactive and adaptive approaches to environmental challenges. Conversely, a fixed mindset can lead to apathy and resistance to change, hindering engagement in sustainable practices. Understanding these mindsets is essential for developing strategies to promote positive attitudes and behaviours towards environmental sustainability.

Recent research has provided valuable insights into the significant impact of growth mindsets on pro-environmental action (Duchi et al., 2020). Through an extensive survey of American adults, the study went beyond mere attitudes towards climate change. It explored beliefs about its mitigation, pro-environmental tendencies, and self-reported engagement in eco-conscious behaviours.

The study revealed that individuals with a growth mindset towards the world showed a greater openness to climate change. This mindset went beyond just attitudes, influencing beliefs about mitigating climate change and leading to an increased inclination towards pro-environmental behaviours.

Furthermore, participants with a growth mindset experienced positive changes in attitudes, beliefs, and behavioural tendencies after being exposed to persuasive and informative discussions on climate change. Additionally, there was a temporal connection, with those holding solid beliefs in the world's adaptability reporting more frequent engagement in pro-environmental actions ten days later. This temporal link highlights the transformative power of a growth mindset, sparking immediate responses and sustaining behavioural changes over time.

These findings carry significant implications. Cultivating a growth mindset is a powerful psychological catalyst, breaking down barriers to pro-environmental action. The adaptive worldview fostered by a growth mindset shapes immediate perceptions. It fosters a long-term commitment to environmentally conscious behaviours. Recognising and promoting such mindsets is crucial for overcoming psychological barriers and promoting positive actions in climate change and environmental sustainability efforts.

Stress mindset

In recent years, a ground-breaking concept known as the "stress mindset" has emerged in stress research. This perspective challenges the traditional belief that stress has a solely negative impact on individuals. Instead, it suggests that individuals' perception of stress profoundly influences their coping mechanisms, impacting their physical and mental wellbeing and outcomes.

Consider a scenario where you receive distressing news about environmental devastation, triggering a stress response. In this moment, your body reacts swiftly. Your liver releases fat and sugar into your bloodstream, providing energy. Meanwhile, your breathing deepens, ensuring a surge of oxygen to your heart, which beats faster, delivering vital resources to your muscles and brain. It's as if your body is mobilising an army, coordinating your faculties to think clearly and respond effectively to the challenge of bad news. Stress, therefore, isn't an enemy but an ally, helping you have a clear mind to come up with solutions to a problem. It enhances your performance, sharpens your response, and improves your behaviour, bringing out your best qualities.

Suppose we hear distressing news and experience stress, and our body and mind are primed to address the challenge but fail to take action. In that case, we can become trapped in a cycle of stress with no outlet. This can lead to prolonged periods of rumination and dwelling on the issue. Hours after hearing the news, we might still find ourselves sitting at the kitchen table, consumed by thoughts of the stressor. Recognising and embracing our stress can motivate us to take constructive action, releasing stress and contributing to environmental efforts.

Embracing a stress-is-enhancing mindset, which views stress as an opportunity for personal growth, correlates with a greater likelihood of employing effective coping strategies than adopting a stress-is-debilitating, negative perspective on stress (Crum et al., 2013).

The stress mindset's adaptability is particularly intriguing (Jamieson et al., 2018). Research suggests that our stress mindset can be modified relatively easily when individuals have a balanced understanding of stress's implications. This discovery highlights the potential of interventions and educational programmes to reshape individuals' attitudes towards stress, promoting enhanced mental and emotional health.

In essence, the emergence of the stress mindset signifies a paradigm shift, emphasising the significant influence of beliefs and attitudes on one's experience of stress. Cultivating a stress-is-enhancing mindset facilitates effective coping strategies. It mitigates the negative consequences of stress, fostering greater resilience and overall wellbeing. As our understanding of stress continues to evolve, interventions targeting stress mindsets may prove invaluable in strengthening individuals' mental and emotional fortitude.

Sustainable consumption involves conscientious choices to reduce environmental, social, and economic impacts. This approach emphasises environmental impact, social responsibility, economic considerations, lifestyle choices, circular economy principles, and consumer awareness. Individuals with a stress-is-helpful mindset are more inclined to adopt sustainable consumption practices (Gorgani, 2020). This connection highlights the importance of considering psychological factors in sustainability research.

Below are examples illustrating how individuals may respond to climate change based on different stress paradigms.

Stress-is-debilitating mindset

Individuals, often unknowingly, may shift their focus away from the root causes of environmental stress, choosing avoidance over active engagement. This avoidance can gradually lead them into a state of denial, where increasing anxiety or sadness becomes intertwined with the unseen effects of climate change.

Instead of addressing the primary sources of their distress, individuals might find temporary relief in activities like mindfulness or deep breathing. While these coping strategies provide immediate comfort, they risk diverting attention from the urgent need to tackle the underlying environmental crises causing their stress, leaving the fundamental issues unaddressed.

In dealing with the psychological burden of climate change, some people may turn to escapism, seeking refuge in alcohol, drugs, or addictive behaviours. This retreat offers a temporary escape from the weight of climate-related stress.

Another response to climate-induced pressure involves redirecting energy and focus away from discussions about climate or topics that amplify mental distress. However, this avoidance carries risks, potentially hindering individuals from actively contributing to sustainable solutions and participating in collective efforts towards positive environmental change.

Understanding how individuals cope with environmental stress is crucial for developing tailored interventions for mental wellbeing. It also empowers individuals to actively promote sustainable practices actively, fostering a resilient community attuned to environmental concerns.

Stress-is-helpful mindset

As individuals confront the challenges of environmental turmoil, they may encounter various stressors. However, those who adopt a mindset that views stress as a catalyst for growth often employ adaptive strategies, highlighting the importance of proactive engagement. Here are vital actions characteristic of individuals embracing such a mindset:

1 **Acknowledging Reality**: The first crucial step is recognising the harsh reality of climate change and facing its associated stressors directly. This acknowledgement lays the foundation for a practical and informed response to urgent environmental issues. By confronting the challenges, individuals equip themselves with the understanding needed to navigate the complex landscape of climate change more effectively.
2 **Developing Strategic Plans:** Individuals with a stress-is-enhancing mindset proactively plan responses to environmental stressors. This involves considering both individual and collective actions to mitigate the impacts of climate change, adopting sustainable practices, and actively contributing to broader conservation initiatives. Strategic planning transforms intentions into concrete steps towards positive change.

3 **Seeking Knowledge and Community:** Understanding the multifaceted nature of climate change, individuals actively seek relevant information, support, or community. This includes staying informed about ongoing environmental efforts, forming alliances with like-minded individuals or organisations, and accessing resources to deepen their commitment to pro-environmental endeavours. Seeking support fosters a sense of communal solidarity and shared responsibility.
4 **Taking Action:** Translating intention into action is essential to effective coping. Individuals with a stress-is-enhancing mindset proactively engage in tangible measures to address, alleviate, or transform environmental stressors. This may involve participating in environmental advocacy, adopting sustainable lifestyle choices, and actively contributing to initiatives promoting ecological wellbeing.
5 **Cultivating Positivity:** Making the best of challenging circumstances requires a perspective that reframes environmental obstacles as opportunities for personal and collective growth. Individuals with such a mindset view these hurdles as pathways to development. This positive reassessment triggers a transformative shift in perspective, promoting resilience and determination in the face of adversity. It empowers individuals to focus their efforts on constructive solutions.

By embracing these adaptive strategies, individuals with a stress-is-enhancing mindset can navigate the challenges posed by climate change and actively contribute to its mitigation, embodying a spirit of resilience and environmental awareness.

Time perspectives and pro-enviro behaviour

Our perception of time plays a silent yet powerful role in shaping our experiences and interactions with the world around us. Take John, a war veteran haunted by his traumatic past, where past horrors overshadow his every moment of joy. In contrast, there's Mary, living in the present without being burdened by past sorrows or future uncertainties, embracing life as it comes.

John and Mary represent extreme examples of how our views on time can impact our wellbeing, a concept explored by Zimbardo and Boyd (2008). They identified five primary time perspectives. Let's review them in the context of pro-environmental behaviour.

1 **Negative Past:** Individuals who dwell on past environmental destruction may feel overwhelmed by anxiety and depression, leading to a lack of motivation to engage in present conservation efforts. For example, someone who witnessed the devastation of a beloved forest due to deforestation might feel too disheartened to participate in reforestation projects.
2 **Positive Past:** Those who have fond memories of engaging with nature in their childhood may be more inclined to support conservation efforts. For

instance, someone who has joyful memories of birdwatching with their grandparents may be motivated to volunteer for local bird habitat restoration projects.

3 **Present Fatalism:** Individuals who feel powerless about the future of biodiversity may struggle to see the point in conservation actions. They may become apathetic or disengaged from environmental efforts because they believe their actions won't make a difference.
4 **Present Hedonistic:** People who prioritise immediate pleasure and enjoyment may not prioritise conservation efforts if they perceive them as sacrificing present enjoyment. For example, someone who continues to use plastic straws or throws baby wipes into the toilet.
5 **Future-oriented:** Future-oriented individuals may dedicate themselves to long-term conservation goals, sometimes at the expense of immediate enjoyment. For example, a scientist focused on biodiversity conservation may spend long hours researching and advocating for policy change, sacrificing leisure time and relationships to pursue their long-term vision for biodiversity protection.

However, focusing too much on any single perspective can be harmful, as John and Mary exemplify. John's fixation on the adverse past risks exacerbates his PTSD. At the same time, Mary's present-focused outlook may lead her to engaging in risky behaviours.

Striking for a balanced perspective is key – a state of wellbeing where present enjoyment, positive memories from the past, and future goals are all valued and considered. Negative past regrets and present fatalism should be avoided. This equilibrium, known as the wellbeing optimum, promotes a fulfilling life.

When promoting pro-environmental behaviour, it's crucial to consider individuals' predominant time perspectives. For those focused on the present, highlighting the immediate pleasures of eco-friendly actions may be more effective than discussing future consequences. Positive psychology interventions, such as fostering gratitude for environmentally conscious actions, can enhance wellbeing and inspire pro-environmental behaviour.

Similarly, for those with a fatalistic present outlook, gently expanding their temporal horizons rather than directly challenging their beliefs may be more effective in promoting change.

Understanding and aligning time perspectives offer a nuanced approach to encouraging positive behavioural change, enriching individuals' engagement with life's complexities.

Crowding out

The concept of "crowding out" plays a significant role, especially in environmental discussions. Take Seamus, for instance, who expressed frustration online about having to switch from plastic to paper straws. Instead of convincing Seamus of the change's collective benefits, a more practical approach is needed. Trying to understand Seamus's preference for pleasure and nostalgic memories associated with plastic straws will shift the conversation towards finding the best reusable straw that satisfies Seamus's enjoyment while aligning with his growing environmental awareness. This type of discussion would make it more relevant for Seamus, making him more likely to contribute to the environmental cause while feeling personally satisfied.

Now, let's look at Marta's recycling habits, which are focused on avoiding penalties. Despite facing fines for improper recycling, Marta continues to put regular waste in recycling bins to avoid fees. Changing Marta's beliefs about the past is difficult, so focusing on the present-day benefits of recycling for herself and the community could be more effective. This reframes the conversation towards positive transformation and communal responsibility, instilling hope and optimism.

Researchers found the link between Zimbardo's temporal perspectives and environmental concerns. They discovered that individuals with positive past and future orientations are more likely to be environmentally conscious. A meta-analysis across seven nations highlights the significant impact of future orientation on pro-environmental attitudes and behaviours, especially among those already inclined towards environmentalism. This emphasises the crucial role of future time perspective in fostering and sustaining pro-environmental actions.

Using time perspective for pro-environmental actions

Time perspectives intricately shape our attitudes and actions towards pro-environmental efforts, sustainability, and responses to climate change. Effective communication in environmental discourse involves customising our message to align with individuals' unique temporal perspectives. Here's how we can navigate different temporal lenses:

Negative Past Perspective

For those burdened by past grievances, our communication should offer hope and highlight the transformative potential for change. We guide them towards present and future possibilities by acknowledging past failures while showcasing success stories and nature's resilience.

Positive Past Perspective

Individuals cherishing positive memories can be engaged by weaving their past accomplishments into the present or future environmental action. By acknowledging their successes in relation to pro-environmental actions, we build a bridge to a greener future.

Present-Hedonistic Perspective

Emphasising the joy and satisfaction found in sustainable living can motivate those focused on present pleasures. From enjoying nature to minimising waste, we help them align with environmental stewardship.

Present-Fatalistic Perspective

We expand their temporal horizons to empower those feeling powerless in the face of environmental challenges. We instil a sense of purpose and agency by illustrating how individual actions contribute to broader environmental outcomes.

Future-Oriented Perspective

Individuals focused on future goals are inspired by narratives that emphasise the importance of pro-environmental actions in safeguarding the planet for future generations. We invite them to join us in creating a greener world by connecting sustainability to global wellbeing.

Understanding these diverse time perspectives allows us to tailor our approach to environmental campaigns, education, and interventions. Recognising the richness of temporal attitudes, we foster a collective response to environmental stewardship, leading to harmonious coexistence with the natural world.

Changing time perspective

Do you need help with your eco-future? This transformative activity guides individuals towards self-discovery and fosters a more environmentally comprehensible world. Backed by research (King, 2001), engaging in the Best Possible Self activity can enhance self-discipline and offer clarity on our aspirations for a sustainable future. Moreover, it can boost optimism (Peters et al., 2013) and life satisfaction (Peters et al., 2013). The positive ripples extend further to elevate self-esteem (Owens & Patterson, 2013), inducing a cascade of positive emotions (Sheldon & Lyubomirsky, 2006; Layous et al., 2013), and improving engagement (Layous et al., 2013).

Best possible eco-self (adapted from King, 2001)

1 Gather recycled paper and an eco-friendly pen.
2 Set an alarm for 20 minutes to immerse yourself in thoughts about your future. In the future, everything will go as well as it could regarding environmental outcomes.
3 Envision the realisation of all your eco-life dreams. Picture yourself having worked hard and succeeded in accomplishing sustainable life goals and adding value to the climate change practice.
4 Write about this imagined Best Possible Eco-Self, capturing the essence of your eco-dreams. Then come up with a step-by-step process on how you can realise your eco-dreams.
5 Make this eco-activity a daily ritual, repeating it for four consecutive days.

Maximising vs satisficing

Pursuing perfection can become a heavy burden in the midst of decisions relating to pro-environmental behaviours. The constant quest for the ideal plan or a flawless climate action can, paradoxically, diminish our subjective wellbeing (Schwartz et al., 2002). This phenomenon is particularly evident in societies with many choices, and the pressure to make the absolute best decision weigh heavy on individuals. Research indicates that adopting an attitude of needing to make the best choice often leads to subsequent regret, impacting our wellbeing negatively (Roets et al., 2012).

For individuals entrenched in the pursuit of perfection, striving for the ideal decision can often lead to strategic failures. This approach assumes we possess complete knowledge of the environment that influences their decision-making process. Similarly, individuals passionate about environmental causes may be pressured to make flawless decisions regarding their impact on the planet. However, the reality remains that we only have access to partial information necessary for making perfect decisions, rendering our choices flawed.

Recognising perfection as an unattainable ideal is important to help us embrace the concept of "good enough" decisions. This shift in perspective can relieve pressure, reduce anxiety, and enhance overall wellbeing (Schwartz, 2004). Recent research suggests that being a maximiser and seeking the best possible outcome may lead to lower performance (Jain et al., 2013). Maximisers often overestimate their performance forecasts, displaying excessive confidence in their predictions. In contrast, satisficers who opt for "good enough" choices without too much deliberation often outperform maximisers in various situations.

When it comes to climate action, let us consider the case of an individual committed to reducing their carbon footprint. They might find themselves overwhelmed by the array of choices available, from eco-friendly products to lifestyle changes for sustainability. Striving for a perfect, carbon-neutral lifestyle could lead to decision paralysis, hindering their journey towards meaningful change. However, by embracing the concept of "good enough" decisions, such as choosing reusable shopping bags or cutting down on meat consumption, they can make tangible contributions to environmental conservation without the weight of perfection.

Similarly, organisations dedicated to climate advocacy may grapple with the pressure to implement flawless sustainability initiatives. They may feel compelled to invest significant resources in complex, high-tech solutions, believing that anything less would adequately address the environmental crisis. However, by embracing the concept of good enough decisions, they can focus on implementing practical, cost-effective measures that promote sustainability within their operations and communities.

Ultimately, the relentless pursuit of perfection in environmental stewardship, as in any other field, can often result in inefficiency and frustration. On the other hand, adopting the idea of "good enough" decisions empowers individuals and organisations to take meaningful action towards environmental conservation, free from the burden of unattainable standards of perfection.

Shifting to satisficing for environmental impact

Helping individuals transition from maximising to satisfying involves cultivating awareness of their maximising tendencies and recognising the negative consequences of this behaviour. Here is a guide to facilitating this shift, particularly in the context of pro-environmental behaviour:

1 **Awareness Building**

- Reflect on your decision-making patterns and recognise if you lean towards maximising.
- List the potential downsides of constantly pursuing the perfect choice, especially in environmental actions.

1 **Highlighting Imperfection**

- Emphasise that perfect decisions are rare, and pursuing perfection may increase stress and anxiety.
- Consider past positive outcomes associated with "good enough" decisions, fostering a sense of accomplishment and wellbeing.

1 **Performance Insights**

- Consider why satisficers often outperform maximisers in forecasting and actual performance.
- Consider why quicker decision-making and acceptance of good enough options may improve overall performance.

1 **Environmental Impact Perspective**

- Apply these principles to environmental choices, encouraging others to prioritise actionable and realistic decisions over exhaustive deliberation.
- Showcase examples of successful environmental initiatives born out of pragmatic decision-making.

In essence, embracing satisficing rather than maximising not only benefits personal wellbeing but also has the potential to enhance our collective impact on the environment. Individuals can navigate their environmental journey more confidently and effectively by understanding that perfection is an unattainable goal and that good enough decisions can pave the way for meaningful change.

Conclusion

This chapter has delved into the profound impact of mindset on various aspects of our lives and its role in shaping our wellbeing. Mindsets, comprising beliefs, attitudes, and perceptions, significantly influence how individuals interpret and respond to the world, impacting their behaviour, emotions, and cognitive processes. Throughout this chapter, we discussed the complexities of passion, recognising its role as a driving force for environmental enthusiasts and acknowledging its potential negative impact on wellbeing. We explored the concept of growth mindset and its potential to enhance pro-environmental actions, shedding light on the importance of adaptive thinking in addressing environmental challenges. Additionally, we examined stress mindset research, highlighting its relevance in assisting individuals experiencing eco-anxiety and stress related to climate change. Lastly, we explored the mindsets associated with maximising and satisfying thinking, emphasising their significance in determining our happiness and satisfaction with actions taken to safeguard the planet. By understanding these diverse mindsets, we gain valuable insights into their influence on our overall wellbeing and our capacity to contribute positively to environmental conservation efforts.

References

Baudon, P., & Jachens, L. (2021). A scoping review of interventions for the treatment of eco-anxiety. *International Journal of Environmental Research and Public Health*, 18(18), 9636. https://doi.org/10.3390/ijerph18189636

Choong, Y.-O., Ng, L.-P., Tee, C.-W., Kuar, L.-S., Teoh, S.-Y., & Chen, I.-C. (2020). Green work climate and pro-environmental behaviour among academics: The mediating role of harmonious environmental passion. *International Journal of Management Studies*, 26(2), 77–97. https://doi.org/10.32890/ijms.26.2.2019.10520

Crum, A. J., Salovey, P., & Achor, S. (2013). Rethinking stress: The role of mindsets in determining the stress response. *Journal of Personality and Social Psychology*, 104(4), 716–733. https://doi.org/10.1037/a0031201

Duchi, L., Lombardi, D., Paas, F., & Loyens, S. M. (2020). How a growth mindset can change the climate: The power of implicit beliefs in influencing people's view and action. *Journal of Environmental Psychology*, 70, 101461. https://doi.org/10.1016/j.jenvp.2020.101461

Dweck, C. S. (1998). The development of early self-conceptions: Their relevance for motivational processes. In J. Heckhausen & C. S. Dweck (Eds.), *Motivation and self-regulation across the life span* (pp. 257–280). Cambridge University Press. https://doi.org/10.1017/CBO9780511527869.012

Jain, K., Bearden, J. N., & Filipowicz, A. (2013). Do maximizers predict better than satisficers? *Journal of Behavioral Decision Making*, 26(1), 41–50. https://doi.org/10.1002/bdm.763

Jamieson, J. P., Crum, A. J., Goyer, J. P., Marotta, M. E., & Akinola, M. (2018). Optimizing stress responses with reappraisal and mindset interventions: An integrated model. *Anxiety, Stress, and Coping*, 31(3), 245–261. https://doi.org/10.1080/10615806.2018.1442615

Junot, A., Paquet, Y., & Martin-Krumm, C. (2017). Passion for outdoor activities and environmental behaviors: A look at emotions related to passionate activities. *Journal of Environmental Psychology*, 53, 177–184. https://doi.org/10.1016/j.jenvp.2017.07.011

King, L. A. (2001). The health benefits of writing about life goals. *Personality and Social Psychology Bulletin*, 27(7), 798–807. https://doi.org/10.1177/0146167201277003

Layous, K., Nelson, S. K., & Lyubomirsky, S. (2013). What is the optimal way to deliver a positive activity intervention? The case of writing about one's best possible selves. *Journal of Happiness Studies*, 14(2), 635–654.

O'Keefe, P. A., Dweck, C. S., & Walton, G. M. (2018). Implicit theories of interest: Finding your passion or developing it? *Psychological Science*, 29(10), 1653–1664. https://doi.org/10.1177/0956797618780643

Owens, R. L., & Patterson, M. M. (2013). Positive psychological interventions for children: A comparison of gratitude and best possible selves approaches. *The Journal of Genetic Psychology: Research and Theory on Human Development*, 174(4), 403–428. https://doi.org/10.1080/00221325.2012.697496

Pennebaker, J., & Evans, J. F. (2014). *Expressive writing: Words that heal*. Idyl Arbor.

Pennebaker, J. W. (1997). Writing about emotional experiences as a therapeutic process. *Psychological Science*, 8(3), 162–166. https://doi.org/10.1111/j.1467-9280.1997.tb00403.x

Pennebaker, J. W., & Chung, C. K. (2011). Expressive writing: Connections to physical and mental health. In H. S. Friedman (Ed.), *Oxford handbook of health psychology* (pp. 417–437). Oxford University Press.

Pennebaker, J. W., Kiecolt-Glaser, J. K., & Glaser, R. (1988). Disclosure of traumas and immune function: Health implications for psychotherapy. *Journal of Consulting and Clinical Psychology*, 56(2), 239–245.

Peters, D. H., Adam, T., Alonge, O., Agyepong, I. A., & Tran, N. (2013). Implementation research: What it is and how to do it. *BMJ*, 347, f6753. https://doi.org/10.1136/bmj.f6753

Roets, A., Schwartz, B., & Guan, Y. (2012). The tyranny of choice: A cross-cultural investigation of maximizing-satisficing effects on well-being. *Judgment and Decision Making*, 7(6), 689–704. https://doi.org/10.1017/S1930297500003247

Schwartz, B. (2004). *The paradox of choice: Why more is less*. HarperCollins Publishers.

Schwartz, B., Ward, A., Monterosso, J., Lyubomirsky, S., White, K., & Lehman, D. R. (2002). Maximizing versus satisficing: Happiness is a matter of choice. *Journal of Personality and Social Psychology*, 83(5), 1178–1197. https://doi.org/10.1037//0022-3514.83.5.1178

Sheldon, K. M., & Lyubomirsky, S. (2006). How to increase and sustain positive emotion: The effects of expressing gratitude and visualizing best possible selves. *The Journal of Positive Psychology*, 1(2), 73–82. https://doi.org/10.1080/17439760500510676

Sloan, R. P., McCreath, H., Tracey, K. J., Sidney, S., Liu, K., & Seeman, T. (2007). RR interval variability is inversely related to inflammatory markers: The CARDIA study. *Molecular Medicine*, 13(3–4), 178–184. https://doi.org/10.2119/2006–00112

Spera, S. P., Buhrfeind, E. D., & Pennebaker, J. W. (1994). Expressive writing and coping with job loss. *Academy of Management Journal*, 37(3), 722–733.

Vallerand, R. J., et al. (2003). Les passions de l'âme: On obsessive and harmonious passion. *Journal of Personality and Social Psychology*, 85(4), 756–767. https://doi.org/10.1037/0022-3514.85.4.756

Zimbardo, P., & Boyd, J. (2008). *The time paradox: The new psychology of time that will change your life*. Free Press.

Part 3

The impact of wellbeing within broader social pro-environmental networks

In the final Part 3 of the book, we shift our focus from individual wellbeing to the broader perspective of social networks to which individuals belong. We assume that these networks either already prioritise environmental behaviour or are receptive to it.

Chapter 8 delves into the familial perspective, examining how pro-environmental actions can serve as a unifying force within families. We explore how such actions can strengthen familial bonds and amplify overall wellbeing.

Moving forward to Chapter 9, we investigate the potential of pro-environmental actions to enhance the wellbeing of individuals within the community. We specifically explore the role of a sense of belonging in fostering wellbeing through environmental engagement.

Finally, in Chapter 10, we extend our outlook to encompass the future, considering the implications of engaging in pro-environmental behaviours for individuals, families, communities, and society. We identify the future of environmental engagement as a promising opportunity to improve overall wellbeing across various levels of social organisation.

DOI: 10.4324/9781003452676-10

8 Family's health and wellbeing

In a cosy suburban neighbourhood, the Smith family, comprising parents Mark and Sarah and their two children, Emily and Jake, embarked on a journey of environmental stewardship that would redefine their family dynamic.

It all began with a simple idea sparked by Emily, passionate about biodiversity. One sunny afternoon, as the family gathered in their back garden for a picnic, Emily shared her concerns about the declining bird population in their area. Her words resonated deeply with the rest of the family, igniting a collective desire to take action.

Determined to make a difference, the Smith family rallied, brainstorming ways to support local biodiversity and promote pro-environmental behaviours. Mark suggested setting up bird feeders and nesting boxes in their garden. At the same time, Sarah proposed planting native flowers to attract pollinators like bees and butterflies.

With enthusiasm, the family eagerly got to work, transforming their garden into a haven for wildlife. Emily and Jake filled the bird feeders and watched visiting birds. At the same time, Mark and Sarah planted various native plants and flowers.

As days turned into weeks, the Smith family's garden became a bustling ecosystem full of full of life. Birds singing, bees buzzing and butterflies dancing made all of them very happy.

But the impact of their actions extended far beyond their garden. Inspired by their success, the Smiths embarked on a mission to spread awareness about the importance of biodiversity and pro-environmental behaviours within their community. They organised neighbourhood clean-up events, participated in local conservation projects, and even started a community garden to promote sustainable food practices.

Through it all, the Smith family served as each other's biggest cheerleaders, supporting and encouraging one another every step of the way.

DOI: 10.4324/9781003452676-11

Whether it was celebrating a new bird species spotted in their garden or Jake's presentation on recycling at school, their bond grew more with each shared accomplishment.

In the end, their journey of environmental stewardship not only transformed their surroundings but also strengthened their familial ties. Together, the Smith family proved that anything is possible with unity, determination, and a shared love for the planet.

Family is a significant catalyst for pro-environmental action. Within this context, values are transmitted, and behaviours are modelled in all family members. Particularly noteworthy is the role of children, who often mirror their parents' environmental values and behaviours, becoming critical players in environmental stewardship. Conversely, in some families, the younger generation spearheads environmental initiatives and advocates for pro-environmental causes.

In this chapter, we delve into the family dynamics that either drive or hinder the pursuit of pro-environmental objectives. We explore the factors influencing collective action towards environmental stewardship by examining how values, motivations, and behaviours intersect within family settings.

Central to our discussion is the introduction of an evidence-informed model. This model serves as both a theoretical framework and a practical tool, empowering young people and their parents to leverage their inherent strengths in biodiversity initiatives. By tapping into their unique qualities, families can discover new avenues for environmental activism and cultivate a sense of purpose and agency in their joint endeavours.

Moreover, we delve into the concept of belonging and its profound impact on engaging families in community-led actions. Through research insights and practical applications, we discuss the transformative power of fostering a sense of belonging within family units. This unity fosters solidarity and collective action, serving as important elements for effective environmental preservation efforts.

Drawing from the voices of the Let It Bee project participants, we showcase first-hand accounts of positive psychology's power in enhancing family wellbeing and galvanising collective action. These narratives illustrate how families can synergise their strengths, resources, and aspirations to protect biodiversity and embrace pro-environmental initiatives.

As we navigate this terrain, we are guided by the belief that by harnessing families' collective power, we can forge a sustainable future where environmental stewardship is woven into the fabric of everyday life. Through our exploration, we aim to inspire and empower families to embark on a journey of shared purpose, resilience, and impact in safeguarding the planet for future generations.

Background to family relationship

The parent-child relationship is a cornerstone of human life, profoundly shaping the development and wellbeing of both parties. Children rely on their parents from birth for nurturing, care, and guidance, laying the groundwork for an enduring and influential bond.

This bond takes shape in infancy through everyday interactions like feeding, comforting, and responding to the child's needs. These early experiences form the basis of attachment, a deep emotional connection that fosters feelings of security and trust. Studies show that secure attachment is linked to positive outcomes such as better emotional regulation, social skills, and resilience later in life (Bakermans-Kranenburg & Van IJzendoorn, 2011).

As children grow, so too does the parent-child relationship, evolving through various stages of development (Laursen & Collins, 2004). During childhood, parents play a vital role in providing structure, instilling values, and serving as role models for their children (Darling & Steinberg, 1993). Effective communication is critical, allowing children to express themselves while parents offer support, guidance, and encouragement.

However, as children enter adolescence, the parent-child relationship changes significantly (Laursen & Collins, 2009). Adolescents may seek greater independence and autonomy, challenging parental authority and asserting individuality. While this transition can lead to tension and conflict, maintaining open lines of communication and mutual respect is crucial for preserving the bond between parent and child.

Despite the challenges that may arise, the parent-child relationship continues to evolve throughout life (Birditt et al., 2010). Even as children become adults and establish their own lives, the bond between parent and child remains deeply rooted. Parents continue offering support, guidance, and unconditional love, while children provide care, respect, and appreciation.

However, despite the inherent strength of the parent-child bond, it is not uncommon for this relationship to experience periods of distance or estrangement (Luescher & Pillemer, 1998). This drift can occur for various reasons, including life transitions, communication breakdowns, unresolved conflicts, and cultural or generational differences.

Life transitions, such as moving away for college or starting a career, can create physical distance between parents and children. At the same time, communication breakdowns can erode trust and intimacy in the relationship. Additionally, unresolved conflicts or past trauma from childhood can contribute to feelings of estrangement between parent and child, as can differences in values or expectations.

Pro-environmental activities and family

Engaging in pro-environmental activities offers more than just a chance to impact the planet positively – it also creates valuable opportunities for parents and children to strengthen their bond. Whether planting trees, cleaning up local parks, or joining community conservation efforts, working together towards a shared goal fosters teamwork, cooperation, and a sense of accomplishment.

However, it is not just about the tasks (Amiot et al., 2007). A shared passion for biodiversity can deepen the connection between parents and children, providing opportunities to explore nature reserves, observe wildlife, or embark on hiking adventures together. These experiences allow families to immerse themselves in the beauty of the natural world and spark meaningful conversations about environmental stewardship and the importance of protecting our planet for future generations.

Moreover, participating in pro-environmental actions instils a sense of purpose and responsibility in children, empowering them to become environmentally conscious citizens (Whitmarsh et al., 2011). When parents model eco-friendly behaviours and instil a love for nature in their children, they lay the foundation for a lifelong commitment to environmental conservation and sustainability.

In essence, finding a mutual interest in pro-environmental actions and biodiversity can bring parents and children closer while fostering a shared sense of purpose and passion for protecting the planet. By nurturing this connection and working together towards a common goal, families strengthen their bonds and make a lasting impact on the environment for generations to come.

The profound bond between parents and children became palpable in the Bee Well pro-environmental initiative (Burke & Corrigan, 2024). Their family connection deepened significantly as they embarked on the mission to care for bees and safeguard the environment to support their survival. Unlike other aspects of farm life that typically failed to capture children's attention, the bees sparked a keen interest and sense of responsibility in them.

Weekends transformed from mundane routines to a time of shared purpose and camaraderie as families devoted themselves to the wellbeing of their buzzing companions. Children eagerly joined their parents in tending to the hives, absorbing knowledge about beekeeping and biodiversity conservation. What initially felt like a chore evolved into a cherished family activity, with Saturdays often dedicated entirely to nurturing the bee colonies and enhancing the surrounding ecosystem.

The impact of the Bee Well project extended far beyond the family farm. Children, inspired by their experiences, shared a newfound passion for bees and environmental stewardship with their broader community. Some young people even sought to deepen their skills and knowledge by joining adult beekeeping clubs, eager to learn from experienced practitioners and contribute to the cause alongside their parents. This ripple effect of the project in the community is a testament to its effectiveness and potential for widespread change.

Family socialisation theory

Families play a crucial role in fostering biodiversity protection and environmental stewardship by educating young people about the importance of change. Schultz (2001) introduced the Family Socialisation Theory, which emphasises the significant influence of family dynamics and parental behaviours on shaping children's attitudes and actions towards the environment. This theory underscores the essential role of families in nurturing environmental awareness and activism among youth.

At the heart of this theory lies the notion that parents serve as primary influencers, transmitting environmental values and behaviours to their children through various channels. Parents profoundly shape their children's environmental consciousness, whether through open discussions about environmental issues or leading by example with actions like recycling and conserving energy. For instance, when children witness their parents actively participating in eco-friendly activities, they are more inclined to adopt similar behaviours.

Moreover, Family Socialisation Theory emphasises the importance of parent-child dialogues surrounding environmental topics. By engaging in meaningful conversations, sharing knowledge about nature, and exploring the outdoors together, families foster a deeper understanding of environmental challenges among children. This heightened awareness often becomes a catalyst for children to participate actively in environmental conservation efforts.

Additionally, the theory recognises the influence of external factors, such as parental education, socio-economic status, and cultural background on family dynamics and environmental socialisation. These broader contextual elements interact with parental practices to shape children's perspectives and actions concerning the environment.

In essence, Family Socialisation Theory highlights the indispensable role of the family environment in nurturing environmental consciousness and activism in children. By understanding the intricate interplay between family dynamics and parental behaviours in shaping children's environmental attitudes, researchers and educators can develop targeted approaches to promote eco-friendly behaviours within families. Ultimately, this fosters the growth of a generation of environmentally aware and engaged individuals.

Children's impact on family's pro-environmental behaviours

Children can play a pivotal role in driving pro-environmental changes within families. Studies exploring how children's environmental activities impact their family's behaviours have provided valuable insights into household environmental awareness dynamics. Research indicates a reciprocal relationship between children's attitudes and behaviours and their family's environmental practices.

For instance, children actively participating in pro-environmental actions like recycling and conserving energy can positively influence their family's environmental behaviour (Schultz & Tabanico, 2007). Furthermore, a longitudinal study conducted by Otto and Pensini (2017) supported these findings. It examined how children's participation in environmental education programmes affected their family's behaviour. The study found that children engaged in such programmes were more likely to influence their family members to adopt pro-environmental behaviours, such as reducing water usage and waste generation. This research highlighted the significant role of children as change agents within their households.

Children can influence their family's change via different avenues. Firstly, they often become advocates for environmental issues within their households, engaging their parents and siblings in conversations about eco-friendly practices. For instance, a child might educate their family about the importance of recycling or conserving energy, initiating discussions that lead to collective efforts to adopt greener habits.

Moreover, children serve as influential role models for their family members. When children consistently demonstrate eco-conscious behaviours like turning off lights or using reusable items, they inspire others in the household to follow suit. Parents and siblings are more likely to model these behaviours, recognising their positive impact on the environment and the importance of aligning with such practices.

Furthermore, children's active participation in environmental activities can motivate their families to become more environmentally conscious. Whether joining community clean-up events, planting trees, or attending educational programs, children's involvement fosters a sense of shared purpose and encourages their families to join similar initiatives. Witnessing their children's passion for environmental causes prompts family members to support and engage in these activities, leading to collective efforts in environmental stewardship.

Additionally, children can uniquely challenge existing norms and habits within their families. By initiating conversations and posing thought-provoking questions about sustainability, they encourage their family members to reflect on their behaviours and consider greener alternatives. This ongoing dialogue promotes a culture of introspection and improvement, driving positive changes in the family's approach to environmental conservation over time.

The influence between children and family members is two-directional. For example, parental modelling can impact children's environmental attitudes and behaviours. Parents who demonstrated pro-environmental behaviours, such as using public transportation or composting, significantly influenced their children's environmental consciousness (Evans et al., 2012). This way, children were likelier to model and incorporate their parents' behaviours into their daily routines.

However, in this chapter, let us delve into how the influence of family members can enhance the environment and lead to higher levels of wellbeing. We will examine the role of character strengths in this journey and how these strengths can be used within families to enhance wellbeing. Subsequently, we will explore how families can employ capitalisation to connect with their members and celebrate each other's successes in pro-environmental actions. This chapter will investigate the concept of savouring as a tool applicable to families. While these concepts can individually improve personal wellbeing, their collective use within a family context can amplify health and wellbeing while fostering pro-environmental behaviour.

Character strengths

In the transformative year of 1998, Professor Martin Seligman assumed the presidency of the American Psychological Association, marking a pivotal moment introducing the new field of positive psychology. In his inaugural address, Seligman encouraged psychologists to shift their focus from pathology to human potential. He called for departure from the conventional Diagnostic and Statistical Manual of Mental Disorders (DSM) used for diagnosing mental illnesses, and aspired to create an "anti-DSM" – a manual devoted to understanding and harnessing what is right with individuals.

Seligman envisioned a compendium that would show human strengths, allowing individuals to understand their innate capabilities and empower themselves to cultivate the finest aspects of their being. This revolutionary concept aimed to redefine the mental health and wellbeing narrative, emphasising a proactive and positive approach to psychology.

Several years later, Seligman's vision came true with the publication of a taxonomy of Character Strengths in 2004, co-authored with Christopher Peterson. Drawing from an exhaustive literature review and insights gained through focus groups involving psychologists, philosophers, sociologists, and other experts, this enduring framework has become a handbook in assessing and understanding human strengths.

The taxonomy of Character Strengths (Table 8.1) is a testament to the transformative power of Seligman's vision. It has shifted the paradigm of psychology, moving from a discipline focused on deficits to one that celebrates and amplifies the inherent goodness within each individual. This shift inspires

Table 8.1 Seven virtues and 24 character strengths

Wisdom: creativity, curiosity, judgment, love of learning, perspective
Courage: bravery, perseverance, honesty, zest
Humanity: love, kindness, social intelligence
Justice: teamwork, fairness, leadership
Temperance: forgiveness, humility, prudence, self-regulation
Transcendence: appreciation of beauty and excellence, gratitude, hope, humour, spirituality

hope and optimism for the future of psychology, a future that embraces the potential for growth and flourishing in every individual.

Like moral traits manifested by individuals (Park et al., 2004), character strengths transcend contexts and lifespans, embody moral values, and are cultivated for their intrinsic worth. These strengths are not only measurable but also universally recognisable across diverse cultures.

Empirical evidence from positive psychology interventions shows the significant impact of applying character strengths, especially signature strengths – in which an individual excels and experiences personal wellbeing when using them (Seligman et al., 2005). Recognising the efficacy of purposefully leveraging one's character strengths as a positive psychological intervention to enhance wellbeing, recent research has extended its focus to the practical application of strengths in diverse settings.

For example, we once worked with a client who disliked his job and wanted to change it. When prompted, he recognised his top character strengths. However, he admitted he did not didn't use them at work. We spent a few weeks helping him bridge a gap between what came naturally to him (character strengths) and what he did for many hours daily. Soon, he began to enjoy his work and re-engaged with it. Two years later, he was still doing the same work and loving it. Research indicates that daily strength use is associated with higher levels of psychological flourishing (Hone et al., 2015). The trick here was to help the client use his strengths.

Extending the application of strengths and values to various life domains demonstrates their potential impact on actions, and this concept holds particular relevance for pro-environmental behaviours. Recognising the pivotal role of character strengths in shaping individual behaviour and exploring their application in fostering environmentally conscious actions is imperative. By integrating positive psychology principles into environmental initiatives, such as workplace sustainability programs or educational curricula, an opportunity exists to harness the intrinsic motivation and wellbeing associated with character strengths to drive positive change. This approach aligns with the broader understanding that cultivating strengths and values can extend beyond personal fulfilment to contribute positively to societal and environmental wellbeing. Table 8.2 provides examples of how this could be accomplished within a family unit.

Character strengths and pro-environmental actions

Of all character strengths, creativity and self-regulation are associated with various pro-environmental and sustainable behaviours in urban residents across Mexico (Corral-Verdugo et al., 2015). Similarly, individuals with character strengths of love, kindness, fairness, leadership, appreciation of beauty and excellence, gratitude, hope, humour, and spirituality display greater engagement in environmental actions (Meyer & Warren, 2021). These findings suggest that emphasising character strengths may contribute

Table 8.2 Applying character strengths to pro-environmental behaviour within a family (adapted from Warren & Coughlan, 2016)

Wisdom and knowledge

- Creativity [Originality, Ingenuity]
  As a family, sitting down, discussing, devising and implementing innovative ways of living and consuming to minimise environmental impact.
- Curiosity [Interest, Novelty-Seeking, Openness to Experience]
  Maintaining an active interest in ongoing environmental experiences; exploring and gaining insights into one's environmental impact; embracing new, environmentally friendly ways of living. Sharing them with the family members and growing their knowledge together.
- Judgment [Open-Mindedness, Critical Thinking]
  Deliberating environmental matters thoroughly together as a family, considering all perspectives, avoiding hasty conclusions, assessing environmental claims, exploring pros and cons of all environmental matters, defining problems, and adapting opinions based on new information, impartially weighing all evidence, including environmental, societal, and social norms.
- Love of Learning
  Mastering new pro-environmental skills, topics, and knowledge systematically, moving beyond curiosity to systematically expand one's understanding. Perhaps, signing up to an environmental group or a club together.
- Perspective [Wisdom]
  As a family, considering and discussing the broader picture and situating pro-environmental behaviours within a larger context.

Courage

- Bravery [Valour]
  Confronting environmental threats, challenges, or opposition with resilience; cheering each other on within a family, advocating for what is right for the environment, even in the face of opposition; acting on convictions, even if unpopular or costly, encompassing physical bravery and beyond. Standing strong for the environment, as a family.
- Persistence [Perseverance, Industriousness]
  As a family, completing pro-environmental tasks despite internal and external obstacles; pushing through challenges to achieve environmental goals; taking pleasure in accomplishing pro-environmental tasks.
- Integrity [Authenticity, Honesty]
  Speaking truthfully about one's pro-environmental behaviour and practices, presenting oneself genuinely, being without pretence, and taking responsibility for feelings and actions towards nature and the environment.
- Zest [Vitality, Enthusiasm, Vigour, Energy]
  Allowing each family member to express a passion for the environment translating into wholehearted pro-environmental behaviours, avoiding half-hearted or indifferent approaches, and feeling alive and activated to enjoy the environment thoroughly.

Humanity

- Love
  Valuing a close relationship with nature, animals, and the planet and seeing it as an extension of a family; establishing connections with animals and plants; recognising the importance of "mother earth."

(*Continued*)

Table 8.2 (Continued)

Kindness [Generosity, Nurturance, Care, Compassion, Altruistic Love, "Niceness"]: Taking action out of a sense of care for nature, coming up with ways to perform acts of kindness, as a family, and conducting environmentally positive deeds, contributing to the improvement of nature.
– Social Intelligence [Emotional Intelligence, Personal Intelligence]: Being aware of the motives and feelings of oneself and other family members and understanding how they match with the desire for pro-environmental behaviour, knowing what actions contribute to being more pro-environmental and recognising one's limitations.
Justice
– Fairness Treating the environment with fairness and justice, just as the family treats each family member, avoiding bias in decisions about nature, and moderating behaviour to give nature a fair chance.
– Leadership Encouraging pro-environmental change within a family, promoting and supporting environmental actions on a larger scale, and striving to see pro-environmental initiatives come to fruition.
– Citizenship [Social Responsibility, Loyalty, Teamwork] Collaborating effectively as a family member dedicated to environmental care, acknowledging everyone's role in protecting the environment, expressing loyalty to the group, fulfilling one's share, participating in activities like signing petitions, making donations, and belonging to environmental groups.
Temperance
– Forgiveness and Mercy Understanding the impact of humanity on the environment and aspiring to do better, as a family unit; forgiving those, including oneself, who have caused harm to the environment; offering second chances; avoiding vengeful attitudes.
– Modesty and Humility Allowing one's pro-environmental accomplishments to speak for themselves, avoiding the spotlight, and refraining from seeing oneself as more special than one is.
– Prudence Exercising caution in choices, avoiding undue risks, refraining from saying or doing things that might later be regretted for the environment, carefully weighing alternatives, and choosing the least damaging option.
– Self-Regulation [Self-Control] Regulating family's consumption formally to minimise waste of natural resources, pollution, and harm to habitats, opting for alternatives even when they don't offer immediate financial benefits.
Transcendence
– Appreciation of Beauty and Excellence [Awe, Wonder, Elevation] Noticing and appreciating the beauty and symmetry of the natural world and enjoying it on outings with the whole family.
– Gratitude Expressing appreciation for nature; being aware of and appreciating positive environmental efforts made by a company, community, institutions, or individuals; dedicating time to express thanks.

(*Continued*)

Table 8.2 (Continued)

- Hope [Optimism, Future-Mindedness, Future Orientation]
  Anticipating positive outcomes for environmental wellbeing in the years to come fosters the belief that personal and collective family efforts can actively shape a flourishing future for nature and the environment.
- Humour [Playfulness]
  Approaching environmental care with a light-hearted perspective, finding amusement in the challenges and consequences of both pro-environmental and non-pro-environmental behaviours expressed through family humour and playful engagement.
- Spirituality [Religiousness, Faith, Purpose]
  Allowing a "green" philosophy to influence the family's choices and actions, understanding one's role in the broader environmental context, and embracing beliefs about the purpose of life that not only guide environmental behaviour but also offer solace and meaning.

Strengths-pro-environmental family contract (adapted from Waters, 2020)

1. Assemble all family members and discuss what actions contribute to environmental wellbeing and fulfilment for your family.
2. Based on your previous discussion, establish two family goals to enhance your collective contribution to environmental conservation.
3. Explore how each family member can leverage their strengths to support the family in achieving the environmental goals.
4. Throughout the following week, actively observe and acknowledge instances where family members demonstrate their strengths in promoting pro-environmental actions.
5. Reconvene at the end of the week to assess your progress towards the family's environmental goals and discuss any challenges or successes you encounter.
6. If you continue this activity, repeat steps 3–5 every week to sustain and enhance your family's commitment to pro-environmental behaviour.

to a heightened sense of personal competence and empowerment, countering the prevailing apathy often observed when individuals perceive their contributions as inconsequential in the face of extensive environmental challenges (Fragnière, 2016). Thus, developing a habit of using character strengths can help incite pro-environmental behaviours.

According to Warren and Coughlan (2016), tourists who used their character strengths demonstrated a higher propensity to engage in pro-environmental behaviours and embrace pro-environmental attitudes. This demonstrates a significant connection between character strengths and inclination towards environmentally responsible actions and attitudes.

Delving deeper into the relationship, specific character strengths emerged as particularly influential in shaping pro-environmental behaviours and attitudes. Among these, traits such as curiosity, judgement, social intelligence, self-regulation, hope, and citizenship were notably prevalent among individuals committed to environmental sustainability.

Curiosity, as a character strength, played a pivotal role in fostering a genuine interest in the natural world. Tourists who were curious were likelier to seek out and appreciate information about local ecosystems, wildlife, and conservation efforts. This innate curiosity increased environmental awareness, prompting individuals to make informed and eco-conscious travel choices.

Judgement, another character strength, empowered tourists to assess their actions' environmental impact critically. Those with a strong sense of judgement were adept at evaluating the consequences of their choices on the local environment, encouraging responsible decision-making. This heightened awareness translated into tangible efforts to minimise negative impacts and support sustainable practices.

Social intelligence emerged as a character strength correlated with pro-environmental behaviours in a communal context. Tourists with vital social intelligence were more likely to collaborate with local communities, engage in conservation initiatives, and contribute positively to the overall wellbeing of the destinations they visited. This emphasised the role of social connections and collaborative efforts in fostering environmentally friendly tourism.

Self-regulation, a character strength linked to discipline and impulse control, was crucial in promoting sustainable behaviours among tourists. Individuals with high levels of self-regulation were better equipped to adhere to environmentally responsible practices, such as waste reduction, energy conservation, and mindful consumption. This self-discipline contributed to a more sustainable and eco-friendly travel approach.

As a character strength, hope proved to be a driving force behind positive environmental attitudes and actions. Tourists who possessed a hopeful outlook were more likely to believe in the efficacy of their contributions to environmental conservation. This optimism fuelled proactive engagement in eco-friendly practices, reflecting a belief in the collective power to impact the environment positively.

Citizenship, as a character strength, was associated with a sense of responsibility and commitment to the broader community, including the global environment. Tourists who exhibited strong citizenship values demonstrated a willingness to participate actively in initiatives promoting environmental sustainability. This broader perspective on citizenship emphasised the interconnectedness of individuals with the global ecosystem, encouraging responsible and conscientious travel behaviours.

In summary, the research highlighted the role of character strengths in influencing tourists' pro-environmental behaviours and attitudes. Traits like curiosity, judgement, social intelligence, self-regulation, hope, and citizenship emerged as influential factors in shaping a more sustainable approach to travel. Recognising and fostering these character strengths can be pivotal in promoting environmentally responsible tourism practices and contributing to the overall wellbeing of the destinations visited.

Capitalisation

For decades, the quality of people's relationships was often assessed based on how they navigated arguments. However, a shifting tide in psychological research now suggests that our daily responses to each other's triumphs and joys wield greater influence over relationship enhancement (Gable et al., 2004). These moments of shared celebration occur far more frequently than contentious debates, presenting ample opportunities for nurturing relational bonds, including those in the context of environmental action.

The crux of our research lies in the concept of 'capitalisation response' – the art of genuinely engaging with and celebrating others' good news with enthusiasm and interest (active-constructive) rather than dismissing it (passive-destructive), critiquing it (active-destructive) or offering lukewarm acknowledgement (passive-constructive) (Gable & Reis, 2010). This practice, driven by positive emotions (Kaczmarek et al., 2022), is a pivotal factor in relationship enhancement.

Capitalisation yields dual benefits: the individual sharing positive news experiences increased wellbeing, while those responding actively and constructively also enjoy rewards, fostering closeness and enhancing relationship satisfaction (Woods et al., 2014; Peters et al., 2018). Therefore, within pro-environmental actions, family members can leverage each other's successes to amplify connection and wellbeing by sharing news of their environmental contributions.

However, it is essential to acknowledge that embracing this relational currency may pose challenges for some. For instance, individuals grappling with social anxiety may find it difficult to both offer and receive capitalisation, potentially leading to a decline in relationship quality within a relatively short period of six months (Kashdan et al., 2013). Therefore, while capitalising on good news can be a powerful catalyst for relationship growth, it necessitates conscious effort and sensitivity, particularly for those navigating psychological barriers. According to Shelley Gable's research on capitalisation, supportive responses after sharing positive events can be categorised into four types:

1 **Active Constructive Response (ACR)**: This response is a powerful tool for fostering positive relationships. It involves actively engaging with the shared positive event in a constructive and enthusiastic manner. The responder shows genuine interest, asks questions, and expresses enthusiasm for the sharer's experience. For example, responding with statements

like "That's wonderful! Tell me more about it" or "I'm so happy for you, let's celebrate!" demonstrates an active and engaged response that can significantly boost the sharer's wellbeing and relationship satisfaction.

2 **Passive Constructive Response (PCR)**: In this type of response, the responder acknowledges the positive event but does so in a more passive or low-key manner. While the response is positive, it needs more active engagement and enthusiasm. For example, saying "That's nice" or "Good for you" without further elaboration or expression of excitement would be considered a passive-constructive response.
3 **Active Destructive Response (ADR)**: This type of response involves engaging with the positive event but in a critical or negative manner. The responder may downplay the significance of the event, offer unsolicited advice, or express jealousy or resentment. For example, responding with statements like "That's not such a big deal" or "You could have done better" would be considered an active, destructive response.
4 **Passive Destructive Response (PDR)**: In this type of response, the responder fails to acknowledge or engage with the positive event and may even dismiss or ignore it altogether. The response is neither supportive nor affirming and may leave the sharer feeling invalidated or ignored. It's important to note that even a seemingly harmless response like silence, changing the subject, or showing disinterest can be considered a passive-destructive response, which can have detrimental effects on relationships and overall wellbeing.

Understanding these four types of responses is critical to navigating positive events in relationships. They play a crucial role in shaping the impact of capitalisation on individuals' wellbeing and relationship satisfaction. Positive and constructive responses, such as active constructive responses, are associated with greater happiness and stronger social bonds, while negative or dismissive responses, such as active, destructive responses, can have detrimental effects on relationships and overall wellbeing. By being aware of these responses, you can actively foster positive relationships and enhance wellbeing.

Gable's research has implications for environmental actions, as it provides deep insights into the power of positive reinforcement and supportive communication and offers practical strategies for igniting sustainable behaviours and family involvement. This research is not just theoretical, but has real-world implications for environmental groups and activists. However, to date, no research was conducted in the context of pro-environmental supports within a family.

At the same time, Gable's research highlights the profound impact of capitalisation within family relations, offering a roadmap for cultivating dynamic and resilient relationships when discussing family members' small and big successes relating to pro-environmental behaviour. By embracing supportive communication and collective celebration principles, the family can harness their members' collective energy, advancing the sustainability agenda and creating a brighter, greener future for all.

Family-based environmental capitalisation (adapted from Lambert et al., 2013)

Encourage each family member to keep a diary for two weeks, documenting their environmental wins. Then, schedule regular family meetings or discussions where members share their personal experiences and successes at least once weekly. This creates a strong sense of family connection and mutual support, where individuals can inspire and uplift one another through their eco-conscious actions, fostering a more profound sense of belonging and camaraderie. By sharing accomplishments and insights, the family reinforces a culture of sustainability and strengthens bonds and connections. Together, they can amplify their impact and drive meaningful change for the environment, instilling in each member a sense of pride and accomplishment.

Savouring

Savouring, a concept described by Bryant and Veroff (2007), embodies the deliberate act of immersing ourselves in and cherishing positive experiences. This notion can be easily introduced as part of pro-environmental actions, where individuals consciously engage with and derive fulfilment from their efforts to safeguard the planet. Savouring encompasses various temporal dimensions – past, present, and future – encouraging individuals to embrace each moment while fully reflecting on its significance.

Reflecting on positive environmental experiences (adapted from Biskas et al., 2022)

Recall a past event involving positive environmental actions or shared experiences. It could be participating in a beach clean-up with friends, planting trees in your community, or celebrating a successful sustainability initiative. Please take a moment to immerse yourself in the memory, recalling the details of the situation and how it made you feel.

Next, distil your recollection into four keywords that encapsulate the essence of your environmental memory. Then, spend 5–10 minutes discussing the specifics of your experience with your family members, from the sights and sounds to the emotions it evoked. Reflect on how this positive environmental encounter contributed to your sense of purpose and connection to nature.

To cultivate the practice of savouring within the context of pro-environmental actions, individuals can adopt specific strategies outlined by Bryant and Veroff. Here are some tailored examples:

1. **Sharing Positive Experiences:** Strengthen bonds with your family by participating in enjoyable activities together, such as beach clean-ups or tree-planting events.
2. **Memory Building:** Taking moments during impactful environmental actions as a family to mentally capture the experience, creating lasting memories to revisit and cherish later, and mentally snapshotting the beauty of a revitalised natural habitat.
3. **Comparing:** Reflecting on the distinctive aspects of family's environmental contributions and appreciating the progress made. Considering the positive influence of eco-conscious choices on the planet and inspiring other family members to follow suit.
4. **Sensory Perceptual Sharpening:** This involves heightening awareness of environmental surroundings by focusing on specific sensory stimuli, such as the chirping of birds in a biodiverse ecosystem or the aroma of fresh air in a preserved wilderness area.
5. **Family Congratulations:** Acknowledging and organising family celebration following contributions to environmental conservation efforts and recognising the impact of eco-conscious decisions, whether through waste reduction or advocacy for sustainable policies.
6. **Absorption:** each family member fully immersing oneself in environmental activities, experiencing a sense of flow and fulfilment while engaging in gardening, hiking, or conservation projects and then discussing with the family members the impact of the absorption on wellbeing.
7. **Behavioural Expression:** Expressing joy and gratitude together with all family members for the natural world through actions like tree planting, supporting local eco-friendly businesses, or participating in environmental activism events.
8. **Temporal Awareness:** Embracing the transient nature of environmental moments, acknowledging the urgency of addressing environmental challenges together while revelling in nature's resilience and renewal.
9. **Counting Blessings:** Cultivating gratitude for the Earth's abundance and biodiversity, expressing appreciation for the interconnectedness of all living beings and ecosystems.
10. **Killjoy Thinking (Reverse):** Abstaining from dwelling on negative thoughts or uncertainties about the environment's future. Instead, focusing on the positive impact of present actions and the potential for collective change.

By integrating these savouring strategies into their environmental endeavours, family members can forge deeper connections with nature, heighten satisfaction with pro-environmental actions, and contribute to a more sustainable future for all.

Delving into the art of savouring brings forth many psychological benefits, as evidenced by recent research findings. Wilson and MacNamara (2021) underscore that embracing the practice of savouring leads to a noteworthy uptick in positive emotions while concurrently mitigating negative ones – a dual effect that Bryant and Veroff (2007) also observed in their seminal work.

Beyond merely influencing emotional states, savouring contributes significantly to what psychologists term "hedonic happiness" – that sense of contentment derived from living a pleasurable life. But the benefits of savouring extend beyond the individual realm; they also permeate our relationships. Lenger and Gordon (2019) found a compelling association between practising savouring and heightened satisfaction within partnerships.

Zooming into the dynamics of couples, the way partners savour experiences, environmental or others, plays a pivotal role in the quality of their relationship. Indulging in the present moment experiences fosters subjective wellbeing among couples, while anticipation savouring – excitedly contemplating future joys – proved to be a potent predictor of relationship satisfaction. This can be done within an environmental context encouraging life partners to connect while living their environmental values.

Moreover, the significance of savouring amplifies during turbulent times within relationships. Samios and Khatri (2019) posit that savouring acts as a robust buffer during periods of stress, enabling couples to navigate challenges together rather than drifting apart. In essence, the practice of savouring emerges as a multifaceted tool for enhancing emotional wellbeing and strengthening the fabric of our relationships.

Relationship savouring for eco-activists (adapted from Borelli et al., 2020)

Consider someone within your family who has provided unwavering support and encouragement for your environmental endeavours. Recollect a specific instance when:

1 They offered compassionate care during a time of environmental concern or frustration.
2 They encouraged you to take a bold environmental action or initiative.
3 They fostered a sense of closeness that made you feel secure, valued, and empowered in your environmental efforts.

Reflect on the details of these interactions, including the circumstances, emotions, and outcomes. Consider how this person's support has strengthened your commitment to environmental activism and deepened your connection with them.

Conclusion

This chapter delved into the profound impact of pro-environmental actions on family connection. It explored how engaging in such actions can bring families closer together and enhance their overall wellbeing. By tapping into their and their family members' character strengths, which are associated with improved wellbeing, families can strengthen their bonds while contributing to biodiversity and pro-environmental efforts. Additionally, savouring pro-environmental moments together and capitalising on each other's successes further deepen familial connections.

Throughout this chapter, we emphasised that pro-environmental actions catalyse families to connect at a deeper level. Furthermore, by leveraging insights from positive psychology science and practice, we can amplify the wellbeing effects of these actions within family dynamics. Families can forge stronger connections through shared values, mutual support, and intentional engagement while making meaningful contributions to environmental conservation.

References

Amiot, C. E., Terry, D. J., Jimmieson, N. L., & Callan, V. J. (2007). A longitudinal investigation of coping processes durinipccg a merger: Implications for job satisfaction and organizational identification. *Journal of Management*, 33(6), 803–830.

Bakermans-Kranenburg, M. J., & van Ijzendoorn, M. H. (2011). Differential susceptibility to rearing environment depending on dopamine-related genes: New evidence and a meta-analysis. *Development and Psychopathology*, 23(1), 39–52. https://doi.org/10.1017/S0954579410000635

Birditt, K. S., Miller, L. M., Fingerman, K. L., & Lefkowitz, E. S. (2010). Tensions in the parent and adult child relationship: Links to solidarity and ambivalence. *Psychology and Aging*, 25(2), 402–411.

Biskas, M., Sirois, F. M., & Webb, T. L. (2022). Using social cognition models to understand why people, such as perfectionists, struggle to respond with self-compassion. *British Journal of Social Psychology*, 61(4), 1160–1182. https://doi.org/10.1111/bjso.12531

Borelli, J. L., et al. (2020). Relational savoring: An attachment-based approach to promoting interpersonal flourishing. *Psychotherapy (Chicago, Ill.)*, 57(3), 340–351. https://doi.org/10.1037/pst0000284

Bryant, F. B., & Veroff, J. (2007). *Savoring: A new model of positive experience*. Lawrence Erlbaum Associates Publishers.

Corral-Verdugo, V., Tapia-Fonllem, C., & Ortiz-Valdez A. (2015). On the relationship between character strengths and sustainable behaviour. *Environment and Behavior*, 47(8), 877–901.

Darling, N., & Steinberg, L. (1993). Parenting style as context: An integrative model. *Psychological Bulletin*, 113(3), 487–496. https://doi.org/10.1037/0033-2909.113.3.487

Evans, G. W., Otto, S., & Kaiser, F. G. (2012). Parental environmental attitudes and their impact on the family. *Environment and Behavior*, 44(6), 677–697.

Fragnière, A. (2016). Climate change and individual duties. *Wires Climate Change*, 7, 798–814. https://doi.org/10.1002/wcc.422

Gable, S. L., Reis, H. T., Impett, E. A., & Asher, E. R. (2004). What do you do when things Go right? The intrapersonal and interpersonal benefits of sharing positive events. *Journal of Personality and Social Psychology*, 87(2), 228–245. https://doi.org/10.1037/0022-3514.87.2.228

Gable, S. L., & Reis, H. T. (2010). Good news! Capitalizing on positive events in an interpersonal context. In M. P. Zanna (Ed.), *Advances in experimental social psychology* (Vol. 42, pp. 195–257). Academic Press. https://doi.org/10.1016/S0065-2601(10)42004-3

Hone, L. C., Jarden, A., Duncan, S., & Schofield, G. M. (2015). Flourishing in New Zealand workers: Associations with lifestyle behaviors, physical health, psychosocial, and work-related indicators. *Journal of Occupational and Environmental Medicine*, 57(9), 973–983. https://doi.org/10.1097/JOM.0000000000000508

Kaczmarek, L. D., et al. (2022). Positive emotions boost enthusiastic responsiveness to capitalization attempts. Dissecting self-report, physiology, and behavior. *Journal of Happiness Studies*, 23, 81–99. https://doi.org/10.1007/s10902-021-00389-y

Kashdan, T. B., Farmer, A. S., Adams, L. M., Ferssizidis, P., McKnight, P. E., & Nezlek, J. B. (2013). Distinguishing healthy adults from people with social anxiety disorder: Evidence for the value of experiential avoidance and positive emotions in everyday social interactions. *Journal of Abnormal Psychology*, 122(3), 645–655. https://doi.org/10.1037/a0032733

Lambert, N. M., Stillman, T. F., Hicks, J. A., Kamble, S., Baumeister, R. F., & Fincham, F. D. (2013). To belong is to matter: Sense of belonging enhances meaning in life. *Personality and Social Psychology Bulletin*, 39(11), 1418–1427. https://doi.org/10.1177/0146167213499186

Laursen, B., & Collins, W. A. (2004). Parent-child communication during adolescence. In A. L. Vangelisti (Ed.), *Handbook of family communication* (pp. 333–348). Lawrence Erlbaum Associates Publishers.

Laursen, B., & Collins, W. A. (2009). Parent–child relationships during adolescence. *Handbook of Adolescent Psychology*, 2, 3–42.

Lenger, K. A., & Gordon, C. L. (2019). To have and to savor: Examining the associations between savoring and relationship satisfaction. *Couple and Family Psychology: Research and Practice*, 8(1), 1–9. https://doi.org/10.1037/cfp0000111

Luescher, K., & Pillemer, K. (1998). Intergenerational ambivalence: A new approach to the study of parent–child relations in later life. *Journal of Marriage and Family*, 60(2), 413–425.

Meyer, S., & Warren M. A. (2021). Exploring the role of character strengths in the endorsement of gender equality and pro-environmental action in the UAE. *Middle East Journal of Positive Psychology*, 7, 65–80.

Otto, S., & Pensini, P. (2017). Nature-based environmental education of children: Environmental knowledge and connectedness to nature, together, are related to ecological behaviour. *Global Environmental Change*, 47, 88–94.

Park, N., Peterson, C., & Seligman, M. E. P. (2004). Strengths of character and well-being. *Journal of Social and Clinical Psychology*, 23(5), 603–619. https://doi.org/10.1521/jscp.23.5.603.50748

Peters, B. J., Reis, H. T., & Gable, S. L. (2018). Making the good even better: A review and theoretical model of interpersonal capitalization. *Social and Personality Psychology Compass*, 12(7), Article e12408. https://doi.org/10.1111/spc3.12407

Samios, C., & Khatri, V. (2019). When times get tough: Savoring and relationship satisfaction in couples coping with a stressful life event. *Anxiety, Stress & Coping: An International Journal*, 32(2), 125–140. https://doi.org/10.1080/10615806.2019.1570804

Schultz, P. W. (2001). The structure of environmental concern: Concern for self, other people, and the biosphere. *Journal of Environmental Psychology*, 21(4), 327–339.

Schultz, P. W., & Tabanico, J. J. (2007). Self, identity, and the natural environment: Exploring implicit connections with nature. *Journal of Applied Social Psychology*, 37(6), 1219–1247.

Seligman, M. E. P., Steen, T. A., Park, N., & Peterson, C. (2005). Positive psychology progress: Empirical validation of Interventions. *American Psychological Association*, 60, 410–421.

Van Lange, P. A., Joireman, J., Parks, C. D., & Van Dijk, E. (2013). The psychology of social dilemmas: A review. *Organizational Behavior and Human Decision Processes*, 120(2), 125–141.

Warren, C., & Coghlan, A. (2016). Using character strength-based activities to design pro-environmental behaviours into the tourist experience. *Anatolia: An International Journal of Tourism & Hospitality Research*, 27(4), 480–492. https://doi.org/10.1080/13032917.2016.1217893

Waters, L. (2020). *Strengh switch: How the new science of strength-based parenting can help your child and your teen to flourish*. Scribe.

Whitmarsh, L., O'Neill, S., & Lorenzoni, I. (2011). Public engagement with climate change: What do we know and where do we go from here? International *Journal of Media & Cultural Politics*, 7(1), 7–25.

Wilson, K. A., & MacNamara, A. (2021). Savor the moment: Willful increase in positive emotion and the persistence of this effect across time. *Psychophysiology*, 58(3), e13754. https://doi.org/10.1111/psyp.13754

Wood, R. G., Moore, Q., Clarkwest, A., & Killewald, A. (2014). The long-term effects of building strong families: A program for unmarried parents. *Journal of Marriage and Family*, 76, 446–463. https://doi.org/10.1111/jomf.12094

9 Community health and wellbeing

Mark tragically lost his wife to illness several years ago, plunging him into a deep sense of helplessness and despair. He felt the world was crumbling around him, with humanity wreaking havoc on nature.

Amidst his depression and existential turmoil, Mark experienced a pivotal moment during a walk with a friend. They discussed the devastation caused by Ash Dieback disease, and Mark began noticing the widespread affliction of ash trees. His friend pointed out that he was only observing death, overlooking the life teeming around them. This conversation sparked a revelation for Mark, prompting him to start paying attention to the natural world beneath his feet.

Mark's friend encouraged him to focus on what he could control, starting with his own space. This conversation had a profound impact on Mark's perspective. As he began to pay attention to his surroundings, he discovered an abundance of herb benedict, a plant used for throat ailments, igniting his curiosity and leading him on a journey of discovery and care.

Learning about the intricate relationship between bees and flowers opened Mark's eyes further. He realised the interconnectedness of all living things and shed the notion that humans were the pinnacle of life. Instead, he marvelled at the intelligence of migratory birds and fell in love with nature's wonders.

Transforming his garden into a habitat for wildlife, Mark witnessed the once-lifeless lawn come alive with the sounds of nature. He recognised the significance of every stage in a plant's life cycle, even in death, in providing sustenance for birds.

Mark's journey was not just for his fulfilment but also for the benefit of future generations. He wanted his grandson to understand the intricacies of life, a knowledge he lacked in his upbringing. Biodiversity became Mark's solace in grieving for his wife, providing him purpose and enjoyment.

Though he remained pessimistic about the planet's future due to society's materialistic focus, Mark found hope in joining like-minded groups dedicated to biodiversity conservation. He emphasised that his actions were not solely for personal enjoyment but stemmed from a deeper understanding of nature's importance.

DOI: 10.4324/9781003452676-12

For Mark, the journey from appreciation to wonder in nature was transformative. He believed that once embarked upon, this journey was irreversible, leading to a profound appreciation for the interconnectedness of all life forms.

Community and environmental practice

The relationship between environmental stewardship and psychological wellbeing is a complex and intricate aspect of community life. In this chapter, we embark on a journey to explore how community-engaged and community-led activities contribute to nurturing psychological wellbeing and fostering meaningful community engagement.

Our exploration is centred around models of engagement that prioritise collaboration and co-production, creating symbiotic partnerships between communities and stakeholders responsible for preserving tangible ecological assets. We explore the profound impacts of managing tangible environmental assets on the health and wellbeing of community members.

Drawing from the unique insights and experiences of the Let It Bee project participants, we share narratives emphasising the transformative power of community-driven initiatives in catalysing positive change. Through their voices, we glimpse the potential of grassroots efforts to cultivate resilience, empowerment, and collective agency.

In navigating this landscape, we aim to give researchers and practitioners actionable insights and effective models of community engagement in environmental management. By showcasing exemplary practices and lessons gleaned, we endeavour to spur a shift towards more inclusive, participatory, and impactful approaches to community-driven environmental action. Throughout our exploration, we interlace our discourse with poignant quotes from project participants, providing first-hand perspectives that imbue our research findings with depth and authenticity.

Ecosystem model and pro-environmental actions

The Wellbeing Ecosystem Model, developed by Carol Ryff and Corey Keyes, provides a holistic framework for understanding and promoting wellbeing on both individual and community levels. This model integrates psychology, sociology, and public health concepts to illustrate how various factors influence wellbeing.

Central to the Wellbeing Ecosystem Model is recognising that wellbeing is shaped by many factors operating across different levels, including individual characteristics, social relationships, community resources, and environmental conditions. These factors interact in complex ways to impact individuals' experiences of wellbeing and quality of life.

The model's domains of wellbeing – self-acceptance, positive relationships, autonomy, environmental mastery, purpose in life, and personal growth – can be applied to understanding individuals' attitudes and behaviours towards the environment.

For instance, *self-acceptance* involves deep connection with nature and recognising its intrinsic value, leading individuals to engage in behaviours that support biodiversity conservation. Feeling a sense of belonging and connection to nature fosters a personal responsibility for its protection.

Positive relationships centred around environmental activism and conservation efforts can motivate individuals to take action to preserve biodiversity. Collaborating with like-minded individuals builds a sense of community and collective responsibility towards environmental stewardship.

Autonomy empowers individuals to make environmentally conscious choices and take initiative in sustainability efforts. Giving people the freedom to decide about eco-friendly practices cultivates a more substantial commitment to pro-environmental behaviour.

Environmental mastery entails developing skills and knowledge related to sustainable living practices, enabling individuals to contribute effectively to biodiversity conservation. Mastering techniques for waste reduction, energy conservation, and habitat protection empowers individuals to make meaningful contributions to environmental sustainability.

Purpose in life emerges from finding meaning and significance in environmental activism and biodiversity conservation. Recognising the importance of preserving biodiversity for future generations instils a sense of purpose and direction in environmental advocacy efforts.

Personal growth involves engaging in environmental education, experiential learning, and reflection, contributing to individuals' growth and development as environmental stewards. Continuous learning and adaptation to environmental challenges foster personal resilience and commitment to conservation efforts.

By applying the principles of the Wellbeing Ecosystem Model to pro-environmental behaviour and biodiversity conservation, communities can create environments that support both human wellbeing and ecological sustainability. Recognising the interconnectedness between individual wellbeing and environmental health is crucial for promoting a holistic approach to conservation that benefits both people and the planet.

Let us now go deeper into each one of the elements in the context of pro-environmental actions and biodiversity.

Self-acceptance

Self-acceptance, the unconditional embrace of one's strengths and weaknesses, is not just about feeling good about oneself; it also supports pro-environmental behaviours. It is particularly important in the context of community as it fosters greater connection between community members. In a landmark study by Ryff and Singer (1996), self-acceptance emerged as a crucial pillar of psychological wellbeing, autonomy, personal growth, and positive relationships. Regarding environmentalism, self-acceptance may empower individuals to acknowledge their ecological impact without harsh self-judgement, fostering a sense of calm and confidence as they navigate sustainability challenges.

Building on this notion, Neff (2003) delved deeper into self-acceptance within the framework of self-compassion theory. Here, self-acceptance becomes a powerful tool for treating ourselves with kindness and understanding, especially in the face of environmental setbacks. By recognising our imperfections and limitations, we're better equipped to confront eco-issues with resilience and empathy, maintaining our commitment to sustainability even when things don't go as planned.

Self-acceptance is a cornerstone of subjective wellbeing (Diener et al., 2010). This means that individuals who genuinely accept themselves are more likely to feel fulfilled and satisfied with life, fuelling their motivation to engage in pro-environmental actions. By embracing their unique identities and recognising their inherent worth, individuals find the confidence to advocate for environmental causes and actively contribute to a greener, more sustainable world.

In a seminal study by Gifford and Nilsson (2014), the researchers explored the intricate relationship between self-acceptance and environmental attitudes among a cohort of Swedish adults. Their findings showed a compelling correlation between self-acceptance, heightened environmental concerns, and pro-environmental attitudes. Remarkably, individuals who exhibited greater self-acceptance were markedly predisposed to expressing genuine care and concern for the preservation and wellbeing of the natural environment.

Expanding upon their research, Gifford and Nilsson discovered intriguing insights into how self-acceptance may serve as a foundational pillar for fostering pro-environmental sentiments within individuals. Their study highlighted that an individual's positive self-regard and acceptance of oneself may profoundly shape their perceptions and attitudes towards the broader ecological landscape. Consequently, those who embraced themselves more fully tended to extend this empathy and concern towards environmental issues, recognising the interconnectedness between personal wellbeing and ecological health.

Moreover, the findings from Gifford and Nilsson's research shed light on the potential implications for environmental advocacy and sustainability initiatives. By cultivating self-acceptance and nurturing positive self-concepts among individuals, a promising avenue exists for nurturing a collective consciousness of environmental stewardship and responsibility. Indeed, promoting self-acceptance may catalyse instigating transformative shifts in attitudes and behaviours towards more eco-conscious lifestyles and decision-making processes.

Corral-Verdugo and his team (2017) examined the intricate relationship between self-acceptance and environmental behaviour among Mexican university students. Their findings revealed that individuals with higher levels of self-acceptance were significantly more likely to engage in pro-environmental actions. From diligently recycling materials to conscientiously conserving energy and opting for eco-friendly transportation, these individuals demonstrated a deep commitment to advancing ecological sustainability.

Building upon these insights, Howell and his colleagues conducted further research in 2013, exploring the influence of self-acceptance on sustainable lifestyle choices among Australian adults. Their study uncovered that individuals with greater self-acceptance were inherently predisposed to embrace eco-friendly behaviours as integral parts of their daily lives. Whether it involved reducing waste, curbing consumption, or advocating for environmental causes, these individuals embodied a holistic ethos of environmental stewardship rooted in self-acceptance.

Moreover, the implications of these studies extend far beyond individual behaviour, resonating within the broader environmental advocacy and sustainability initiatives. By cultivating a culture of self-acceptance, communities can shift towards more sustainable practices and attitudes, fostering a collective commitment to preserving the planet for generations to come.

Furthermore, who embraced higher self-acceptance also reported a stronger bond with the natural world (Nisbet et al., 2009). The study suggested that accepting ourselves more fully may naturally develop a deeper appreciation and reverence for nature.

Building on these results, research further substantiated the initial findings, indicating that those who harboured greater self-acceptance were more likely to view themselves as intricately connected to nature (Zelenski & Nisbet, 2014). These individuals demonstrated a heightened propensity to engage in behaviours that nurture ecological wellbeing, highlighting the potential ripple effect of self-acceptance on pro-environmental actions.

These studies highlight the significant influence of self-acceptance on our relationship with nature and our inclination towards pro-environmental behaviours. By fostering a positive self-regard and acceptance culture, communities can cultivate a collective sense of environmental stewardship, paving the way for concerted efforts to preserve and protect the natural world for generations to come.

Self-acceptance fosters unity and solidarity among members of environmental conservation groups, strengthening their commitment to shared environmental goals. For instance, members of a community-based reforestation project may embrace their diverse backgrounds and skills, working harmoniously to restore degraded ecosystems.

Groups that practice self-acceptance are more effective in collaborating on pro-environmental initiatives. For example, a community recycling program thrives when members accept each other's ideas and contributions, fostering cooperation in waste reduction efforts.

Environmental groups facing external criticism or challenges benefit from a strong sense of collective identity and worth. When a conservation organisation maintains self-acceptance despite criticism from detractors, it remains resilient and focused on its mission to protect biodiversity and ecosystems.

Embracing self-acceptance contributes to a positive group identity around environmental stewardship. For instance, a wildlife conservation group may celebrate its members' diverse perspectives and experiences, fostering a culture of inclusivity and dedication to protecting endangered species.

Here are examples of what community groups can do to foster group self-acceptance:

1 Environmental groups can promote self-acceptance by fostering open communication and dialogue. For instance, a marine conservation team could encourage members to share ideas and concerns openly, creating a supportive environment for collaboration and problem-solving.
2 Recognising and celebrating diversity within environmental groups can strengthen self-acceptance. A biodiversity conservation organisation may highlight members' unique contributions from different cultural backgrounds, fostering a sense of belonging and appreciation for individual perspectives.
3 Environmental groups could establish norms of support and empathy to cultivate self-acceptance. For example, a community garden collective could encourage members to offer assistance and encouragement to one another, nurturing a culture of mutual respect and care.
4 Resolving conflicts constructively could reinforce self-acceptance within environmental groups. When disagreements arise over conservation strategies, group members engage in respectful dialogue and compromise, demonstrating their commitment to collective wellbeing and environmental goals.

Group self-acceptance is fundamental to the success of pro-environmental and biodiversity actions, fostering members' cohesion, collaboration, and resilience. By promoting open communication, celebrating diversity, and establishing support norms, environmental groups create environments where individuals feel valued and empowered to contribute to collective efforts in protecting the planet's ecosystems and biodiversity.

Positive relationships

Community positive relationships, integral to Ryff's model of psychological wellbeing, embody the quality of social bonds and interactions within a community, characterised by mutual respect, trust, support, and cooperation (Ryff & Singer, 1998). These relationships contribute significantly to individuals' overall wellbeing and sense of belonging within their community, shaping mental and emotional health.

Research has consistently shown that positive relationships within the community serve as a crucial source of social support, offering emotional, instrumental, and informational assistance during challenging times (Thoits, 2011). Whether providing a listening ear during difficult circumstances or offering practical help in times of need, the support network within a community enhances individuals' resilience and promotes their psychological wellbeing.

Strong community relationships foster a profound sense of belonging and connectedness among residents, contributing to their overall life satisfaction and happiness (Jetten et al., 2012). Feeling accepted and valued within the community bolsters individuals' self-esteem and promotes a positive sense of identity, reinforcing their commitment to the community's wellbeing.

Positive relationships are the bedrock for community collaboration and collective action. When individuals trust and respect one another, they are more inclined to work together towards common goals, such as environmental conservation and sustainability (Haslam et al., 2010). Shared values and a sense of camaraderie facilitate collective efforts to address pressing environmental challenges effectively.

Positive relations are crucial for cultivating trust and cooperation within the community, which is essential for effective pro-environmental initiatives. When community members trust one another, they are more likely to collaborate on environmental projects, such as organising tree-planting drives or advocating for eco-friendly policies (Bonaiuto et al., 2016).

Open communication and dialogue foster positive relationships and enable community members to address environmental concerns collectively. Creating platforms for discussion and idea exchange allows communities to develop innovative solutions to environmental challenges, ensuring inclusive participation and buy-in from all stakeholders (Díaz et al., 2018).

Embracing diversity could strengthen community relationships and foster a more inclusive approach to pro-environmental actions. Valuing diverse perspectives and experiences enhances empathy and understanding, leading to more comprehensive and equitable environmental initiatives that address the needs of all community members (Schultz et al., 2011).

Mutual respect and empathy are foundational to positive relationships and vital for navigating differences in opinions or approaches within the community. By practising active listening and showing empathy towards differing viewpoints, community members can find common ground and collaborate effectively on pro-environmental goals (Steg & Perlaviciute, 2015).

Acknowledging and celebrating environmental conservation efforts reinforces positive relationships within the community. Recognising individuals' contributions fosters a sense of appreciation and motivates continued engagement in pro-environmental behaviours, strengthening community bonds and collective resolve (Ryan & Deci, 2000).

Positive community relationships foster collaboration, support, and a sense of community belonging. By prioritising mutual respect, trust, and cooperation, community members can work together to promote environmental sustainability and address pressing environmental challenges effectively, ensuring a healthier and more resilient future for all.

Autonomy

Autonomy refers to an individual's ability to act independently, make self-directed choices, and regulate their behaviour by personal values and objectives (Ryff & Keyes, 1995). In community settings, autonomy is pivotal in empowering individuals to engage actively in communal endeavours, make informed decisions, and contribute meaningfully to collective wellbeing. It also refers to the autonomy of a community, which can make decisions and self-directed choices for the benefit of the community members.

Autonomy empowers community members to participate actively in decision-making processes and initiatives. When residents have the autonomy to voice their opinions in local governance meetings or shape community policies, it enhances their sense of belonging and fosters investment in communal welfare.

Autonomy fosters respect for diversity within communities by recognising and valuing individuals' unique perspectives and identities. Communities prioritising autonomy create inclusive environments where diverse voices are heard and respected, promoting equity and cohesion among residents.

Autonomy encourages personal growth within communities by providing opportunities for individuals to pursue their interests and aspirations independently. For example, community members may autonomously engage in environmental education programs or conservation projects, fostering personal growth and contributing to community resilience.

Autonomy encourages community members to take initiative and drive communal initiatives. When individuals autonomously organise events like neighbourhood clean-up campaigns or initiate environmental advocacy projects, it cultivates a sense of ownership and collective responsibility for community wellbeing.

Autonomy fosters meaningful relationships within communities by creating environments where individuals feel respected and valued. When community members have the autonomy to express themselves authentically and engage in relationships based on mutual respect, it strengthens social bonds and promotes community cohesion.

Here are a few examples of community-based autonomous pro-environmental actions:

1 Empowering residents to participate in decision-making processes related to environmental policies or urban planning ensures that community initiatives align with their needs and preferences, promoting autonomy and enhancing communal wellbeing.

2 Supporting autonomy-driven community initiatives, such as establishing community gardens or organising eco-friendly events, fosters a sense of ownership and pride among residents, promoting community cohesion and wellbeing.
3 Creating autonomy-supportive environments within community organisations and workplaces enables individuals to pursue their interests and contribute to shared goals autonomously, promoting wellbeing and community flourishing.

Autonomy promotes community wellbeing by empowering individuals to engage actively in communal life, respect diversity, foster personal growth, take initiative, and cultivate supportive relationships. By prioritising autonomy within communities, stakeholders can create environments conducive to individual flourishing and collective prosperity.

Environmental mastery

Environmental mastery (Ryff & Keyes, 1995) is defined as an individual's capacity to navigate and adapt to their surroundings effectively. It encompasses feeling competent in managing daily life's demands, exerting control over one's environment, and actively shaping it to align with personal needs and aspirations.

In the context of community pro-environmental action, environmental mastery is pivotal in empowering individuals and communities to confront environmental challenges adeptly. This multifaceted construct contributes to community wellbeing by fostering various dimensions of engagement, resilience, and agency, as outlined below.

Individuals with elevated levels of environmental mastery demonstrate proficiency in identifying and resolving environmental issues within their communities. They exhibit proactive behaviour by devising innovative strategies to tackle challenges, such as implementing waste reduction programs, advocating for renewable energy initiatives, or spearheading local conservation efforts.

Environmental mastery entails responsible resource management, which encompasses the judicious use of natural resources and the adoption of sustainable practices. Communities with a mastery orientation prioritise conservation endeavours, including water and energy conservation, sustainable land management, and habitat preservation, thereby minimising ecological footprints and fostering long-term sustainability.

Environmental mastery cultivates a culture of community engagement and collaboration surrounding environmental issues. Individuals who feel empowered and competent to effect change within their environment are more likely to actively participate in community-based initiatives, such as volunteering for conservation projects, engaging in environmental advocacy, or organising community clean-up events.

Communities characterised by high levels of environmental mastery demonstrate resilience in the face of environmental adversities. Individuals adept at environmental mastery exhibit resilience when confronted with natural disasters, climate change impacts, or pollution incidents. They leverage their problem-solving skills and resourcefulness to mitigate risks, recover from setbacks, and foster sustainable practices conducive to community wellbeing.

Environmental mastery instils empowerment and agency among individuals and communities, enabling them to assume control over their environmental destiny. It imbues individuals with confidence in their capacity to drive positive change, fostering advocacy for environmental policy reforms, challenging unsustainable practices, and advocating for environmental justice.

Environmental mastery is a cornerstone of community pro-environmental action, equipping individuals and communities with the skills, resources, and mindset to navigate environmental challenges effectively. By fostering environmental mastery within communities, policymakers, educators, and community leaders can empower individuals to become proactive stewards of their environment, driving meaningful change towards sustainability and ecological resilience.

Purpose in life

Purpose in life embodies a profound sense of meaning, direction, and fulfilment derived from pursuing goals that resonate with the collective values and beliefs of the community members (Ryff & Singer, 1998). When individuals within a community harbour a strong sense of purpose, they are driven and motivated to positively impact the environment they share, guiding their actions towards contributing to causes they collectively deem significant.

The potent catalyst of purpose in life propels pro-environmental action within a community. Let's explore how purpose in life intersects with community-based pro-environmental action.

Grounded in collective purpose, community members are spurred to effect meaningful change in their shared environment. This intrinsic motivation extends to environmental causes, motivating community members to advocate for and engage in actions to preserve local ecosystems and mitigate environmental degradation (Steger & Dik, 2010).

Purpose in life instils communities with a sense of vision and foresight, prompting them to prioritise the wellbeing of future generations and the sustainability of their local environment. Communities with a strong sense of purpose are inclined to adopt sustainable practices and advocate for policies safeguarding the environment for posterity (Steger et al., 2008).

Armed with a collective sense of purpose, communities exhibit resilience when confronted with obstacles on their journey towards environmental conservation. They persevere in the face of setbacks, drawing upon their shared motivation and unwavering commitment to continue striving towards a greener and more sustainable future (Burrow & Hill, 2011).

Purpose-driven communities are inherently inclined to seek connections with each other and forge alliances in pursuit of shared environmental goals. In the domain of environmental activism, communities with a strong sense of purpose gravitate towards collective initiatives, collaborating with local organisations and authorities to effect change and address pressing environmental concerns (Hicks & King, 2009).

Engaging in pro-environmental action imbues community members with a profound sense of fulfilment and wellbeing, aligning with their collective purpose. Contributing to environmental causes evokes feelings of accomplishment and satisfaction, reinforcing the community's sense of purpose and bolstering their collective psychological wellbeing (Kjell et al., 2016).

In essence, purpose in life emerges as a driving force behind a community's commitment to pro-environmental action, imbuing their collective efforts with meaning and significance. By harnessing their shared sense of purpose, communities can forge a path towards a more sustainable future where environmental stewardship and community wellbeing intertwine to create a thriving and resilient local environment for generations to come.

Personal growth

Personal growth encapsulates an individual's journey of continuous development and self-realisation through experiences that foster self-improvement, learning, and resilience (Ryff & Singer, 1998). It entails a process of self-awareness, adaptation, and growth, leading to a deeper understanding of oneself and the surrounding world.

In the context of pro-environmental actions and biodiversity conservation efforts, personal growth is pivotal in empowering individuals to enact meaningful change towards environmental sustainability. Let us explore how personal growth intertwines with pro-environmental actions and biodiversity conservation.

Personal growth nurtures self-awareness and introspection, enabling individuals to recognise their connection to the environment. Through introspection and reflection, individuals develop a heightened sensitivity to their ecological footprint and the significance of biodiversity conservation.

Embracing personal growth involves a commitment to continuous learning and intellectual curiosity. Individuals inclined towards personal growth actively seek opportunities for environmental education, deepening their understanding of ecological principles and conservation strategies essential for effective biodiversity conservation (Stevenson et al., 2013).

Personal growth fosters adaptability and resilience, crucial traits in navigating environmental challenges. Individuals on a path of personal growth view setbacks as opportunities for learning and growth, adapting their strategies to overcome obstacles encountered in their pursuit of biodiversity conservation (Wamsler et al., 2018).

Personal growth instils empowerment and self-efficacy, bolstering individuals' confidence in their ability to effect positive change. As individuals engage in pro-environmental actions, personal growth enhances their belief in their capacity to contribute meaningfully to the preservation of natural habitats and biodiversity (Steg et al., 2014).

Personal growth deepens individuals' connection to nature and fosters a profound appreciation for biodiversity. Through personal growth, individuals derive a sense of purpose and meaning from their efforts to conserve ecosystems and safeguard species diversity, reinforcing their commitment to environmental stewardship (Howell & Allen, 2017).

Personal growth promotes collaboration and collective action among individuals united by a shared commitment to environmental sustainability. As individuals undergo personal growth, they become more inclined to collaborate with others, amplifying the impact of community-based initiatives aimed at biodiversity conservation and ecosystem preservation (Thaler et al., 2018).

Personal growth catalyses fostering environmental consciousness, empowerment, and collaboration among individuals engaged in pro-environmental actions and biodiversity conservation efforts. By prioritising personal growth, individuals can unlock their potential to contribute meaningfully to protecting natural ecosystems and promoting a sustainable future for all species.

Personal Growth Initiative is a positive psychological concept akin to the community wellbeing element of personal growth. It is rooted in proactive efforts towards self-improvement and development and extends its relevance beyond personal domains into community-driven pro-environmental and biodiversity actions. It involves individuals taking proactive steps to expand their knowledge, skills, and experiences, thereby contributing to their growth and the advancement of environmental causes.

Community members begin their growth journey by fostering self-awareness, including understanding their connection to the environment. Activities such as community nature walks or participation in local environmental education programs deepen this connection, enhancing community members' appreciation of ecological systems and biodiversity.

Community-driven personal growth initiatives involve setting environmental goals that resonate with shared values and priorities. For instance, collectively reducing the community's carbon footprint by promoting public transport or adopting community-wide sustainable energy practices showcases proactive environmental action aligned with community values.

Communities committed to personal growth initiatives prioritise lifelong learning about environmental issues. This may involve organising workshops on conservation biology or sustainable agriculture, which provide community members with opportunities to broaden their knowledge base and adapt to evolving environmental challenges.

Community-driven personal growth initiatives thrive on feedback from environmental experts and collaborative efforts among community members.

Engaging in community-based conservation projects or joining local environmental advocacy groups amplifies the impact of individual actions, fostering a collective approach towards environmental stewardship.

Encouraging community members to leave their comfort zones and confront environmental challenges is integral to personal growth initiatives. Initiatives like organising tree-planting events in unfamiliar terrains or advocating for environmental policies amidst opposition cultivate resilience and a willingness to embrace discomfort within the community.

Here are some ideas that can help you engage in community-based Personal Growth Initiatives in the context of pro-environmental behaviours.

1 Engaging in community-based citizen science projects, such as monitoring local bird populations or assessing water quality in nearby streams, enables community members to contribute to environmental research while enhancing their scientific observation skills.
2 Collaboratively advocating for local or national environmental policies empowers community members to voice their concerns and influence decision-making processes. Community-led environmental advocacy campaigns are prime examples of personal growth initiatives driving collective action.
3 Initiating community-based green businesses or launching eco-friendly initiatives, such as establishing zero-waste markets or promoting sustainable fashion, showcases the intersection of entrepreneurial ambition and environmental stewardship within the community.

Personal Growth Initiatives drive community members' commitment to environmental action, facilitating their individual growth and contributing to protecting and preserving biodiversity and ecosystems. Through collective engagement and continuous self-improvement, communities can play a significant role in addressing environmental challenges and fostering a sustainable future for generations to come, thereby enhancing community wellbeing.

As a branch of psychology, Positive Community Psychology emphasises the enhancement of wellbeing, resilience, and flourishing within communities by leveraging strengths-based approaches and collaborative efforts (Kaczynski et al., 2014). When applied to pro-environmental behaviour, positive community psychology underscores the significance of using communities' collective strengths and resources to tackle environmental challenges and promote sustainability.

Positive community psychology contributes to pro-environmental behaviour through its strengths-based approach, which identifies and mobilises unique community assets to promote sustainable practices and protect natural environments (Marques et al., 2018). For example, communities may use local knowledge, cultural traditions, and innovative solutions to address environmental issues such as waste management, pollution, or habitat restoration.

Empowerment and community engagement are pivotal in applying positive community psychology to promote pro-environmental behaviour (Steg & Perlaviciute, 2015). When community members feel empowered to take action and participate in decision-making processes related to environmental issues, they are more inclined to adopt sustainable practices. Positive community psychology advocates for participatory approaches that involve community members in all stages of environmental initiatives, fostering a sense of ownership and commitment to sustainability.

Moreover, social support networks and connectedness are essential in promoting pro-environmental behaviour within communities (Evans et al., 2017). These networks provide individuals with encouragement, resources, and opportunities for collaboration around environmental initiatives. Positive community psychology strengthens social bonds and collective efficacy in driving sustainability efforts by creating supportive environments where community members can share knowledge, experiences, and inspiration related to environmental conservation.

Positive community psychology also recognises the importance of promoting positive emotions and wellbeing to encourage pro-environmental behaviour (Capaldi et al., 2014). Activities that evoke positive emotions such as awe, gratitude, and connection with nature can enhance motivation to engage in environmental conservation efforts. Community-based initiatives such as community gardens, nature walks, and environmental education programs promote wellbeing and cultivate a deeper appreciation for the natural world and a sense of responsibility towards environmental stewardship.

Finally, collaborative problem-solving and innovation are essential to positive community psychology's approach to pro-environmental behaviour (Carmi et al., 2019). By fostering a culture of innovation and resilience, positive community psychology encourages communities to work together to identify creative solutions, pilot new initiatives, and adapt sustainable practices to local contexts, thereby effectively addressing environmental challenges.

In conclusion, positive community psychology offers a holistic approach to promoting pro-environmental behaviour by leveraging community strengths, fostering empowerment and engagement, nurturing social support networks, promoting positive emotions and wellbeing, and encouraging collaborative problem-solving and innovation. By embracing these principles, communities can build a sustainable and resilient future for both people and the planet.

Positivity resonance – emotional contagion

Positivity resonance encapsulates the essence of shared positive emotions and their profound impact on interpersonal relationships (Fredrickson, 2013). At its core, positivity resonance refers to the dynamic and reciprocal exchange of positive emotions between individuals, fostering a sense of connection, understanding, and emotional synchrony within a community.

Fredrickson's broaden-and-build theory of positive emotions is the foundation for understanding positivity resonance mechanisms (Fredrickson, 2001). According to this theory, experiencing positive emotions such as joy, gratitude, and love can broaden individuals' cognitive and behavioural repertoires, increasing psychological resilience, creativity, and social connectedness. This broadening effect enhances individuals' capacity to adapt and thrive and facilitates the formation of meaningful social bonds.

Fredrickson's work highlights the transformative power of positive emotions in shaping social interactions and relationships. By cultivating moments of positivity resonance in everyday interactions, individuals can enhance the quality of their connections with others and foster a greater sense of happiness, resilience, and fulfilment in their lives. It offers valuable insights into the importance of cultivating positive emotions to foster social connection and enhance overall wellbeing. By prioritising positivity and nurturing moments of emotional resonance with others, individuals can build stronger, more fulfilling relationships and lead happier lives.

In environmental activism, positivity resonance emerges as a potent force, igniting a sense of camaraderie and shared purpose among community members while bolstering their commitment to protecting the planet. To begin, fostering positivity and resonance within these communities bolsters social bonds and cooperation. Through collaborative endeavours like group clean-up initiatives, nature excursions, or engaging discussions, individuals forge deeper connections with one another. These collective moments of positivity lay the groundwork for trust and solidarity, amplifying the effectiveness of teamwork and united action in advancing environmental causes.

Moreover, positivity resonance is an inspiration, propelling individuals to embrace pro-environmental behaviours and advocacy. Encounters with nature often evoke emotions like awe, gratitude, and joy, nurturing a profound connection to the natural world and a shared determination to safeguard it. By nurturing these positive sentiments through collective experiences and shared stories, community members feel empowered and compelled to enact meaningful change in the face of environmental challenges.

Furthermore, positivity resonance contributes to the resilience and wellbeing of individuals within these communities. Emotional bonds and supportive relationships provide solace and fortitude during challenging times, preventing burnout and fostering a sense of belonging. Shared encounters of positivity serve as a source of encouragement and renewal, empowering individuals to persevere in their environmental endeavours for the long haul.

In practical terms, environmental communities can cultivate positivity resonance by fostering inclusive and supportive environments that prioritise positive emotions and interpersonal connections. This might involve organising nature outings, hosting storytelling sessions, facilitating group meditation or mindfulness practices, and establishing peer support networks. By deliberately nurturing positivity resonance, environmental advocates and organisations foster a collective sense of purpose and agency, paving the way for more impactful and sustainable action to protect our planet.

Sense of belonging

Sense of belonging among environmental activists is a complex psychological phenomenon that draws upon various theoretical frameworks to understand why individuals are motivated to engage in environmental activism and how their sense of belonging influences their commitment to environmental causes. Several theories in positive psychology provide valuable insights into this topic, shedding light on the underlying mechanisms and processes involved.

The drive to belong is deeply ingrained in human nature, compelling us to seek acceptance and connection within our social circles. According to Baumeister and Leary (1995), this need for belonging is a powerful motivator for our behaviours, guiding us to integrate into communities where we feel accepted and valued.

Recent research shows the profound influence of our innate desire to belong on our participation in environmental conservation efforts. Individuals who prioritise their reputation and yearn for social belonging are more likely to engage in actions that contribute to environmental conservation (Mac Donald & Staats, 2022). This inclination may arise from their community's aspiration to be acknowledged and validated for their eco-friendly behaviours.

Moreover, the link between conservation behaviour and belongingness runs deeper than just seeking recognition. Actively participating in conservation initiatives can foster a sense of camaraderie and shared purpose among individuals, strengthening their bond with like-minded peers and creating a sense of belonging within their social circles.

As conservation practices become ingrained in the norms and values of communities, individuals may experience a profound connection to their environment and a heightened sense of affiliation with their community. By collectively striving towards shared environmental goals, individuals can further fortify their sense of belonging and purpose within their community, fostering a sense of hope and inspiration.

Ultimately, exploring the intricate interplay between pro-environmental behaviour and the human need for belonging is crucial. By comprehending how engagement in conservation efforts enhances individuals' sense of integration within their communities, we can devise strategies to promote sustainable behaviours while fostering a greater sense of community cohesion and connectedness. Through this exploration, we aim to contribute to initiatives that harness the inherent desire for belongingness to drive positive environmental outcomes and enhance community wellbeing, instilling a sense of motivation and encouragement.

Positive psychology offers several theories that provide insights into a sense of belonging. These theories focus on understanding the psychological factors contributing to individuals' sense of connection, acceptance, and inclusion within social groups or communities. Here are some prominent theories in positive psychology related to a sense of belonging.

Social Identity Theory

Social Identity Theory (SIT), coined by Henri Tajfel and John Turner in 1979, sheds light on the inner workings of community-driven pro-environmental endeavours. At its core, SIT suggests that a significant part of our sense of self and self-worth is rooted in the social groups we belong to. These groups, known as in-groups, offer us a sense of connection, shared identity, and social unity. Conversely, individuals view those outside their group, termed out-groups, as different and often less favourable.

When applied to community pro-environmental action, SIT highlights the significance of group affiliation and collective identity in inspiring individuals to partake in environmental initiatives. As individuals align themselves with a community of environmental activists, they cultivate a unified sense of purpose, shared values, and common aspirations centred around environmental conservation. This shared identity fosters feelings of belonging and camaraderie among group members, motivating them to collaborate towards a shared environmental mission.

Furthermore, SIT highlights the role of social categorisation and comparison in shaping individuals' attitudes and behaviours within the environmental activist community. By identifying with the environmental activist group, individuals categorise themselves as part of this collective entity, distinguishing themselves from those who do not share their environmental values. This process of social categorisation reinforces individuals' dedication to pro-environmental action by solidifying their sense of group identity and solidarity with fellow activists.

Moreover, SIT suggests that group membership can influence individuals' attitudes, beliefs, and behaviours through social influence and conformity. Within the environmental activist community, individuals are exposed to shared norms, values, and behavioural expectations that guide their actions towards environmental stewardship. As group members adhere to these norms and expectations, they contribute to the collective effort to advance environmental sustainability and address ecological challenges.

In essence, SIT provides a framework for understanding the psychological underpinnings of community-led pro-environmental action. By acknowledging the importance of collective identity, social categorisation, and shared norms, environmental activists can harness these principles to foster community, cohesion, and collaborative efficacy within their groups. By bolstering individuals' connection to the environmental movement and nurturing a shared sense of purpose, communities of environmental activists can mobilise collective action and drive positive change towards a more sustainable future.

Conservation of Resources (COR) theory, developed by Hobfoll in 1989, sheds light on the dynamics of pro-environmental activism within communities. This theory delves into how individuals strive to acquire, retain, and safeguard personal and social resources crucial to their wellbeing. Among these resources, a deep sense of belonging emerges as a psychological asset bolsters individuals' resilience and adaptive coping strategies.

COR theory highlights the importance of cultivating a robust sense of belonging among community members in environmental advocacy. When individuals feel connected to supportive social networks and environmental activist communities, they gain access to essential resources that empower them to navigate challenges and uphold their psychological wellbeing. This sense of belonging creates a nurturing environment where individuals can share experiences, exchange ideas, and collaborate on collective environmental initiatives.

Moreover, COR theory highlights the protective role of belongingness against stress and adversity within environmental activist groups. By fostering a sense of community and social support, these groups buffer against the negative impacts of environmental challenges and setbacks. When individuals feel valued and supported within their activist communities, they are better equipped to cope with stressors and persevere in their efforts to safeguard the environment.

Additionally, COR theory emphasises the significance of collective action in resource conservation. As members of environmental activist communities unite towards common goals, they combine their resources, expertise, and efforts to tackle environmental issues effectively. Through collaborative endeavours, these communities can enact meaningful change and advocate for sustainable practices that benefit the environment and society.

The COR theory offers valuable insights into the psychological mechanisms underlying pro-environmental activism within communities. By recognising the pivotal role of belongingness as a fundamental psychological resource, environmental activist groups can foster community, resilience, and collective efficacy among their members. By fostering supportive social networks and promoting collaborative efforts, these communities can effectively preserve resources and drive positive environmental change for the betterment of current and future generations.

Integrated model of group membership and social identity

In recent years, one particular theory that gained traction was the Integrated Model of Group Membership and Social Identity (IMGMSI), introduced by Jetten, Haslam, and Cruwys in 2017.

Building upon established theories like SIT and SCT, the IMGMSI offers a more nuanced perspective on how individuals within environmental activist circles perceive themselves amidst group dynamics. This model recognises the intricate nature of group membership, acknowledging that activists often belong to multiple environmental groups with distinct values, norms, and identities.

According to the IMGMSI, an environmental activist's sense of belonging is shaped by personal identity (how they view themselves as individuals) and social identity (how they identify within activist groups). At the heart of this theory lies the notion that activists find belonging through their alignment with environmental causes and the extent to which they feel embraced, respected, and supported by their fellow activists.

Moreover, the IMGMSI highlights the significance of group dynamics, such as cohesion and social support, in moulding activists' sense of belonging. It underscores the importance of interpersonal connections and shared experiences within activist communities as crucial elements in fostering a sense of belonging.

Furthermore, the IMGMSI acknowledges the complexity of belongingness within environmental activism, recognising that individuals may experience varying levels of belongingness across different activist groups and contexts. This theory also highlights the fluid nature of belongingness, suggesting that it can evolve in response to shifts in group dynamics, life events, and environmental challenges.

The Integrated Model of Group Membership and Social Identity provides valuable insights into the sense of belonging among environmental activists. Considering individual and social factors, this framework offers a comprehensive understanding of how activists perceive themselves within their communities and the implications for their wellbeing and collective efforts towards environmental causes.

Third Culture Kids

Third Culture Kids (TCKs) are a unique demographic group that spends a significant portion of their developmental years in cultures different from their parents' or their passport country's culture. This lifestyle exposes them to diverse environmental attitudes, practices, and challenges across various cultural contexts. As such, community pro-environmental activism can profoundly impact TCKs, shaping their environmental awareness, values, and behaviours.

TCKs often form deep connections with the diverse communities they encounter throughout their upbringing. Immersed in multicultural environments, they develop friendships and bonds with individuals from various backgrounds, fostering a sense of camaraderie and inclusivity. This exposure to different cultures and perspectives enriches their worldview and instills a broad understanding of global citizenship.

However, despite their adaptability and openness to new experiences, TCKs may need help establishing a sense of rootedness or permanence in any cultural context. Constantly moving between countries and cultures, they may feel like perpetual outsiders or cultural chameleons, adept at blending in but never fully belonging. This transient lifestyle can lead to feelings of displacement or identity ambiguity as TCKs navigate the tension between their diverse cultural influences and their desire for a sense of home.

In the context of belonging to environmental communities, TCKs often find solace and connection in shared values and causes. Engaging in pro-environmental activism can give them a sense of purpose and belonging as they collaborate with like-minded individuals to address global environmental challenges. Environmental communities offer TCKs a space where their multicultural backgrounds are celebrated rather than seen as a barrier to belonging, fostering a sense of identity and camaraderie grounded in shared environmental values.

Due to their multicultural upbringing, research suggests that TCKs often develop a deep appreciation for environmental diversity and sustainability. Exposure to different ecosystems, conservation efforts, and environmental policies in various host countries can instil in them a sense of global citizenship and responsibility towards the planet (O'Reilly, 2006). Additionally, many environmental activist communities' multicultural and inclusive nature resonates with TCKs' experiences of belonging to multiple cultures, fostering a sense of identity and connection within these groups (Wells, 2013).

Furthermore, community pro-environmental activism provides TCKs with opportunities for meaningful engagement and action on environmental issues. Participating in local conservation projects, environmental education initiatives, and advocacy campaigns allows TCKs to leverage their diverse cultural perspectives and experiences to effect positive change (Szkudlarek & Usunier, 2016). By collaborating with peers from different cultural backgrounds, TCKs can contribute their unique insights and skills to address environmental challenges innovatively.

However, the transient nature of the TCK lifestyle may also present challenges for their involvement in community pro-environmental activism. Frequent relocations and cultural transitions can disrupt their continuity of engagement with environmental initiatives, making it challenging to establish long-term commitments or connections within local activist communities. Additionally, TCKs may grapple with feelings of environmental rootlessness or identity ambiguity as they navigate shifting environmental values and priorities across different cultural contexts (van der Kolk & Choi, 2019).

In conclusion, community pro-environmental activism plays a significant role in shaping the environmental attitudes, values, and behaviours of TCKs. By providing opportunities for engagement, collaboration, and identity formation, environmental activist communities offer TCKs a platform to contribute positively to environmental conservation efforts while navigating the complexities of their multicultural upbringing. Understanding and supporting TCKs' unique needs and perspectives within environmental activism contexts can enhance their sense of belonging and efficacy as global citizens committed to environmental stewardship.

Forgiveness

Forgiveness is not a simple act of excusing or condoning harmful actions. It is a profoundly personal journey, especially when we find ourselves in the aftermath of conflict or trauma. As Worthington (2006) explains, forgiveness operates on two distinct planes: decisional and emotional, which are deeply intertwined with our personal experiences and emotions.

The decisional aspect involves a deliberate choice to refrain from seeking revenge or retaliation against those who have wronged us. Instead, it is about consciously opting for kindness and pro-social behaviour towards the perpetrator while protecting us from further harm.

Conversely, emotional forgiveness delves into the realm of deep healing. It entails transforming our negative emotions and thoughts associated with the offence into positive ones. Rather than succumbing to bitterness, shame, or anger, we cultivate feelings of compassion, hope, and kindness. This shift not only brings inner peace but also liberates us from resentment.

It is crucial to understand that forgiveness primarily benefits the forgiver, not the wrongdoer. By embracing forgiveness, we free ourselves from negativity and open the door to psychological and physical benefits, paving the way for our personal growth and wellbeing.

The research highlights the toll of harbouring grudges, with a lack of forgiveness correlating with higher levels of depression (Myung-Sun, 2016). While forgiveness may require time and effort (Worthington et al., 2000), it empowers us to reframe our perspective and move forward positively (Ascioglu Onal et al., 2017).

Moreover, forgiveness significantly enhances overall wellbeing, as demonstrated by numerous studies (Bono et al., 2008; Akhtar & Barlow, 2018). It is a potent remedy to stress, reducing heart rate, blood pressure, anxiety, and anger (Subkoviak et al., 1995; Vanoyen et al., 2001), allowing us to shed the weight of negative memories. Ultimately, forgiveness is not just about letting go; it's a profound act of self-care that propels us towards healing and growth, showing us the path to a better, more fulfilling life.

In environmental conservation, forgiveness takes centre stage as a crucial element in navigating the aftermath of ecological devastation. From pollution to habitat destruction, environmental harm encompasses a broad spectrum of actions that degrade ecosystems and jeopardise the wellbeing of communities reliant on these natural resources. Forgiveness in this context entails acknowledging the environmental damage inflicted and its ripple effects on affected communities. It requires releasing resentment and anger towards those responsible and opting instead for empathy and reconciliation.

In addition to improving the health and wellbeing of individuals, another compelling reason for embracing forgiveness in the face of environmental harm is its potential to ignite productive dialogue and collaboration. By extending forgiveness, individuals and communities can unite to address the root causes of ecological degradation and work towards sustainable solutions.

Moreover, forgiveness holds a promising potential for facilitating the healing and restoration of damaged ecosystems and strained human relationships. Just as ecosystems have the remarkable resilience to regenerate with proper care, so can bonds between individuals and communities mend through forgiveness and collective action, instilling a sense of hope and optimism for a better future.

Moreover, forgiveness is a catalyst for breaking the cycle of retaliation and conflict that often exacerbates environmental degradation. Rather than seeking vengeance, forgiveness paves the way for reconciliation and cooperation, empowering individuals and communities to lay the groundwork for transformative change.

It is crucial to note that forgiveness does not absolve responsibility or negate the need for accountability. Holding individuals and entities accountable remains essential for preventing future harm and promoting environmental justice, providing a sense of reassurance and confidence in the process.

In essence, forgiveness is a potent tool for addressing environmental challenges and fostering healing, reconciliation, and collaboration. By embracing forgiveness, we cultivate a deeper sense of empathy and collective responsibility for safeguarding our planet for future generations.

Ecological forgiveness letter (adapted from Worthington et al., 2000)

Over 30 minutes, compose a letter to a party responsible for ecological harm. In this letter:

a Briefly recount the event that caused environmental damage.
b Offer your perspective on the motives behind the offender's actions, seeking to understand their perspective.
c Share your reasons for wanting to extend forgiveness, emphasising the importance of healing and fostering environmental stewardship.
d Clearly state your forgiveness towards the individual or entity responsible for the ecological offence.

This exercise provides an opportunity to reflect on the interconnectedness of human actions and their impact on biodiversity and environmental wellbeing. We can strive towards reconciliation and collective efforts to preserve and protect our planet's precious ecosystems through forgiveness.

Conclusion

In this chapter, we delved into the intricate dynamics of community wellbeing from the perspective of individuals as community members. Beginning with examining the ecosystem model, we highlighted the environmental impact of communities on individuals, particularly in the context of climate change. We then explored the concept of belongingness alongside key theories, clarifying its significance in driving pro-environmental actions and fostering communal cohesion. Additionally, we investigated the transformative potential of positive resonance moments in fostering deep connections within communities. Finally, we discussed the role of forgiveness as a catalyst for strengthening community bonds. Through these explorations, we have gained valuable insights into the multifaceted nature of community wellbeing and the pivotal role of individuals within community contexts.

References

Akhtar, S., & Barlow, J. (2018). Forgiveness therapy for the promotion of mental well-being: A systematic review and meta-analysis. *Trauma, Violence, & Abuse*, 19(1), 107–122.

Ascioglu Onal, A., & Yalcın, İ. (2017). Forgiveness of others and self-forgiveness: The predictive role of cognitive distortions, empathy, and rumination. *Eurasian Journal of Educational Research*, 17(68), 97–120.

Baumeister, R. F., & Leary, M. R. (1995). The need to belong: Desire for interpersonal attachments as a fundamental human motivation. *Psychological Bulletin*, 117(3), 497–529. https://doi.org/10.1037/0033-2909.117.3.497

Bonaiuto, M., Alves, S., De Dominicis, S., & Petruccelli, I. (2016). Place attachment and natural hazard risk: Research review and agenda. *Journal of Environmental Psychology*, 48, 33–53.

Bono, G., McCullough, M. E., & Root, L. M. (2008). Forgiveness, feeling connected to others, and well-being: Two longitudinal studies. *Personality and Social Psychology Bulletin*, 34(2), 182–195.

Burrow, A. L., & Hill, P. L. (2011). Derailed by diversity? Purpose buffers the relationship between ethnic composition on trains and passenger negative mood. *Personality and Individual Differences*, 50(8), 1302–1307.

Capaldi, C. A., Dopko, R. L., & Zelenski, J. M. (2014). The relationship between nature connectedness and happiness: A meta-analysis. *Frontiers in Psychology*, 5, 92737.

Carmi, N., Arnon, S., Orion, N., & Hofstein, A. (2019). Design principles for effective science and environmental education learning environments. *Studies in Science Education*, 55(2), 169–202.

Corral-Verdugo, V., Caso-Niebla, J., Tapia-Fonllem, C., & Frías-Armenta, M. (2017). Consideration of immediate and future consequences in accepting and responding to anthropogenic climate change. *Psychology*, 8(10), 1519–1531.

Díaz, S., et al. (2018). The IPBES Conceptual Framework – connecting nature and people. *Current Opinion in Environmental Sustainability*, 29, 87–96.

Diener, E., et al. (2010). New well-being measures: Short scales to assess flourishing and positive and negative feelings. *Social Indicators Research*, 97, 143–156. https://doi.org/10.1007/s11205-009-9493-y

Evans, L., Maio, G. R., Corner, A., Hodgetts, C. J., Ahmed, S., & Hahn, U. (2017). Self-interest and pro-environmental behaviour. *Nature Climate Change*, 3(2), 122–125.

Fredrickson, B. L. (2001). The role of positive emotions in positive psychology: The broaden-and-build theory of positive emotions. *American Psychologist*, 56(3), 218.

Fredrickson, B. L. (2013). Positive emotions broaden and build. *Advances in Experimental Social Psychology*, 47, 1–53. Academic Press.

Gifford, R., & Nilsson, A. (2014). Personal and social factors that influence pro-environmental concern and behaviour: A review. *International Journal of Psychology: Journal International de psychologie*, 49(3), 141–157. https://doi.org/10.1002/ijop.12034

Haslam, S. A., Jetten, J., Postmes, T., & Haslam, C. (2010). Social identity, health and well-being: An emerging agenda for applied psychology. *Applied Psychology*, 59(3), 379–405.

Hicks, J. A., & King, L. A. (2009). Positive mood and social relatedness as information about meaning in life. *The Journal of Positive Psychology*, 4(6), 471–482.

Hobfoll, S. E. (1989). Conservation of resources: A new attempt at conceptualizing stress. *American Psychologist*, 44(3), 513.

Howell, R. A. (2013). It's not (just) "the environment, stupid!" Values, motivations, and routes to engagement of people adopting lower-carbon lifestyles. *Global Environmental Change*, 23(1), 281–290.

Howell, R., & Allen, S. (2017). People and planet: Values, motivations and formative influences of individuals acting to mitigate climate change. *Environmental Values*, 26(2), 131–155.

Jetten, J., Haslam, C., & Haslam, S. A. (2012). *The social cure: Identity, health and well-being*. Psychology Press.

Kaczynski, A. T., et al. (2014). Are park proximity and park features related to park use and park-based physical activity among adults? Variations by multiple socio-demographic characteristics. *International Journal of Behavioral Nutrition and Physical Activity*, 11, 146.

Kjell, O. N. E., Diener, E., & Suh, E. M. (2016). Ongoing subjective well-being, aspirations, and goals predict increases in extrinsic aspirations among adolescents. *Journal of Personality*, 84(3), 277–289.

Mac Donald, S., & Staats, H. (2022). Conservation as integration: Desire to belong as motivation for environmental conservation. *Society & Natural Resources*, 35(1), 75–91.

Marques, J., Delicado, A., & Gomes, C. M. (2018). Promoting positive reactions toward diversity in a context of economic crisis: The role of positive community psychology. *Journal of Community Psychology*, 46(5), 614–629.

Myung-Sun, K. (2016). Interrelations among forgiveness, depression, stress and resilience in individuals with divorce experience. *Personality and Individual Differences*, 102, 111–116.

Neff, K. D. (2003). Self-compassion: An alternative conceptualization of a healthy attitude toward oneself. *Self and Identity*, 2(2), 85–101.

Nisbet, E. K., Zelenski, J. M., & Murphy, S. A. (2009). The nature relatedness scale: Linking individuals' connection with nature to environmental concern and behavior. *Environment and Behavior*, 41(5), 715–740. https://doi.org/10.1177/0013916508318748

Ryan, R. M., & Deci, E. L. (2000). Self-determination theory and the facilitation of intrinsic motivation, social development, and well-being. *American Psychologist*, 55(1), 68–78.

Ryff, C. D., & Keyes, C. L. (1995). The structure of psychological well-being revisited. *Journal of Personality and Social Psychology*, 69(4), 719–727.

Ryff, C. D., & Singer, B. (1996). Psychological well-being: Meaning, measurement, and implications for psychotherapy research. *Psychotherapy and Psychosomatics*, 65(1), 14–23.

Ryff, C. D. & Singer, B. (1998). The contours of positive human health. *Psychological Inquiry*, 9, 1.

Schultz, P. W., Nolan, J. M., Cialdini, R. B., Goldstein, N. J., & Griskevicius, V. (2011). The constructive, destructive, and reconstructive power of social norms. *Psychological Science*, 18(5), 429–434.

Steg, L., Bolderdijk, J. W., Keizer, K., & Perlaviciute, G. (2014). An integrated framework for encouraging pro-environmental behaviour: The role of values, situational factors and goals. *Journal of Environmental Psychology*, 38, 104–115.

Steg, L., & Perlaviciute, G. (2015). Understanding the human dimensions of a sustainable energy transition. *Frontiers in Psychology*, 6, 144983. https://doi.org/10.3389/fpsyg.2015.00805

Steger, M. F., & Dik, B. J. (2010). Work as meaning: Individual and organizational benefits of engaging in meaningful work. In P. A. Linley, S. Harrington, & N. Garcea (Eds.), *Oxford handbook of positive psychology and work* (pp. 131–142). Oxford University Press.

Steger, M. F., Kashdan, T. B., Sullivan, B. A., & Lorentz, D. (2008). Understanding the search for meaning in life: Personality, cognitive style, and the dynamic between seeking and experiencing meaning. *Journal of Personality*, 76(2), 199–228.

Stevenson, K. T., Peterson, M. N., Bondell, H. D., Mertig, A. G., & Moore, S. E. (2013). Environmental, institutional, and demographic predictors of environmental literacy among middle school children. *PloS One*, 8(3), e59519.

Subkoviak, M. J., Enright, R. D., Wu, C. R., Gassin, E. A., Freedman, S., Olson, L. M., & Sarinopoulos, S. (1995). Measuring interpersonal forgiveness in late adolescence and middle adulthood. *Journal of Adolescence*, 18, 641–655.

Szkudlarek, B., & Usunier, J. C. (2016). TCKs as global boundary spanners: The role of social capital. *Journal of Global Mobility*, 4(1), 71–89.

Thaler, T., et al. (2018). The impact of a participatory approach to environmental education: A quasi-experimental study in 12 German cities. *Environmental Education Research*, 24(7), 1005–1025.

Thoits, P. A. (2011). Mechanisms linking social ties and support to physical and mental health. *Journal of Health and Social Behavior*, 52(2), 145–161.

van der Kolk, H., & Choi, J. (2019). What's home for TCKs? A grounded theory approach to understanding the concept of home in third culture kids. *Frontiers in Psychology*, 10, 609.

Vanoyen, K., Frijda, N., & Zeelenberg, M. (2001). Plausible self-preventing emotions. *Cognition & Emotion*, 15(2), 259–278.

Wamsler, C., Brossmann, J., Hendersson, H., Kristjansdottir, R., McDonald, C., & Scarampi, P. (2018). Mindfulness in sustainability science, practice, and teaching. *Sustainability Science*, 13, 143–162.

Worthington Jr, E. L. (2006). *Forgiveness and reconciliation: Theory and application.* Routledge.

Worthington Jr, E. L., Kurusu, T. A., Collins, W., Berry, J. W., Ripley, J. S., & Baier, S. N. (2000). Forgiving usually takes time: A lesson learned by studying interventions to promote forgiveness. *Journal of Psychology and Theology*, 28(1), 3–20.

Zelenski, J. M., & Nisbet, E. K. (2014). Happiness and feeling connected: The distinct role of nature relatedness. *Environment and Behavior*, 46(1), 3–23. https://doi.org/10.1177/0013916512451901

10 The future of pro-environmental actions, health and wellbeing

In pursuing a healthier, more sustainable world, it is crucial to acknowledge the profound interconnectedness between environmental wellbeing and mental health. This final chapter aims to highlights the synergies between environmental goals and the objectives the World Health Organization set forth to mitigate mental illness and enhance mental wellbeing. By delving into the intersection of these two crucial disciplines, we aim to uncover how individuals, families, and communities can harmoniously engage in pro-environmental actions. This fosters a healthier planet and bolsters their mental health, empowering them to make a significant difference.

Throughout this exploration, we will discern the common ground shared by environmental stewardship and mental wellness, illuminating the potential for mutual reinforcement. By identifying overlaps and complementary strategies, we endeavour to pave a path towards sustainable solutions that benefit the environment and human wellbeing.

Furthermore, this chapter serves as a platform for collective examination of the existing research gaps within the ecological and mental health domains. We aim to catalyse future inquiries by pinpointing these gaps and inspire researchers specialising in ecology and wellbeing to embark on novel investigations. Through our collaborative efforts and innovative approaches, we aspire to bridge these divides and propel forward a holistic understanding of the intricate relationship between environmental sustainability and mental health.

Character strengths

As we venture into the realm of environmental conservation and the protection of biodiversity, it becomes increasingly apparent that our approaches must evolve, embracing the full spectrum of human potential to effect meaningful change. This is where the exploration of Values In Action (VIA) character strengths holds immense promise.

Imagine a world where each individual's unique strengths, whether it be kindness, perseverance, creativity, or love of learning, are harnessed in the pursuit of environmental sustainability. Consider the impact of using

DOI: 10.4324/9781003452676-13

courage to challenge the status quo, temperance to promote mindful consumption, or gratitude to cultivate a deeper connection to the natural world. These character strengths, among many others, have the power to not only inspire individual action but also to catalyse collective movements towards a more harmonious relationship with our planet.

Future research and practice in this area are pivotal. By delving into the nuances of how VIA character strengths can be effectively applied in pro-environmental actions, we unlock a treasure trove of possibilities. For instance, studies may explore how the strength of curiosity can drive innovation in sustainable technologies, how the virtue of humility can foster collaboration in conservation efforts, or how the strength of spirituality can deepen our reverence for the interconnectedness of all life forms.

Moreover, as we strive to protect and improve biodiversity, understanding the role of character strengths becomes even more crucial. Picture communities coming together, each individual drawing upon their inherent strengths to restore ecosystems, preserve endangered species, and advocate for environmental justice. From the resilience of communities rebuilding after natural disasters to the compassion of individuals advocating for the rights of marginalised species, VIA's character strengths offer a roadmap for transformative action.

Indeed, our past research has demonstrated the transformative potential of pro-environmental actions in facilitating the expression and cultivation of individuals' strengths, propelling them towards their best selves. Building upon this foundation, there lies a vast opportunity to further amplify the reach and impact of such actions, potentially heralding a new era of unprecedented human wellbeing.

By expanding our exploration into the intersection of pro-environmental actions and personal growth, we unlock the possibility of not only making these actions more prevalent but also enhancing their profound impact on human flourishing. Imagine a world where every act of environmental stewardship serves not only to protect our planet but also to nurture the holistic wellbeing of individuals and communities.

In this imagined future, engaging in pro-environmental actions becomes synonymous with embarking on a journey of self-discovery, empowerment, and fulfilment. Each recycling effort, tree-planting initiative, or advocacy campaign becomes an opportunity for individuals to tap into their innate strengths, cultivate resilience in the face of adversity, and experience a deeper sense of purpose and connection with the world around them.

Moreover, as the ripple effects of these actions spread throughout society, the collective wellbeing of humanity is able to reach unprecedented heights. Communities united by a shared commitment to environmental sustainability find themselves strengthened by bonds of solidarity and collaboration. Individuals, empowered by their contributions to the greater good, experience heightened levels of happiness, satisfaction, and meaning in their lives.

By further expanding our research in this area, we not only explore the symbiotic relationship between pro-environmental actions and human well-being but also pave the way for transformative change on a global scale. Through innovative interventions, community-based initiatives, and policy advocacy, we have the opportunity to harness the full potential of pro-environmental actions as catalysts for personal growth, societal resilience, and planetary healing.

The importance of future research and practice can be leveraged to address the pressing environmental challenges of our time. By embracing the diversity of human strengths and perspectives, we not only safeguard the future of our planet but also nurture a deeper sense of purpose, fulfilment, and interconnectedness within ourselves and with the world around us. So, let us embark on this journey of discovery, armed with the wisdom of positive psychology and the commitment to forge a brighter, more sustainable future for generations to come.

Individuals and communities pulling together in shared goal

Imagine a world where the threat of environmental crisis is no longer met with fear but with a collective sense of optimism, unity, and purpose. Picture a global community coming together, not in despair but in determination, viewing the challenges ahead not as insurmountable obstacles but as opportunities for growth, collaboration, and innovation.

In this future, people from all walks of life focus on building a better, more sustainable world for future generations. Regardless of background or circumstance, every individual finds themselves inspired by the possibility of creating positive change. They are motivated by the knowledge that their actions, no matter how small, contribute to a more significant collective effort.

Gone are the days of divisive rhetoric and apathy towards environmental stewardship. Instead, conversations are fuelled by hope, creativity, and a shared responsibility towards our planet and its inhabitants. Communities come alive with a renewed sense of purpose as neighbours join hands to plant community gardens, clean up local parks, and implement renewable energy initiatives.

In this transformed future, the journey towards a sustainable future has challenges. Still, each obstacle is met with resilience, perseverance, and an unwavering belief in the power of collective action. Small moments of accomplishment, whether it's witnessing the first sprout of a seedling or celebrating the successful implementation of a water protection programme, are cherished as victories for the planet and humanity.

As individuals come together to tackle environmental challenges head-on, they discover a newfound sense of connection, belonging, and fulfilment. The journey towards sustainability becomes not just a destination but a way of life – a shared adventure filled with moments of joy, camaraderie, and profound personal growth.

So, let us embrace this vision of a world where the future is not a source of fear but an opportunity to create a more sustainable tomorrow. Let us recognise the immense potential within each of us to make a difference, and let us inspire and uplift one another as we embark on this transformative journey together. For in the face of adversity, there is opportunity. And together, united by a common purpose, we have the power to shape a future that is not only sustainable but genuinely extraordinary.

Health and environmental agendas combined

In the new era, imagine a future where the boundaries between environmental stewardship and human wellbeing blur, bringing people together towards progress and transformation. In this vision, two government agendas converge with unprecedented success – one dedicated to nurturing the health of our planet and another committed to improving health and wellbeing of our people.

At the heart of this convergence lies a profound realisation that the fate of our environment and the wellbeing of humanity are inextricably linked. We no longer view these agendas as separate and distinct but as interconnected. As such, governments worldwide unite in an audaciouos and visionary mission to create a future where the health of our planet and our people flourish hand in hand.

There are plenty of initiatives to combat climate change, preserve biodiversity, and safeguard our natural resources. From reforestation efforts to renewable energy transitions and ocean conservation to sustainable agriculture, and the world is full of progress and possibility. Each policy and programme is designed to protect our planet for future generations and enhance the quality of life for all who call it home.

Simultaneously, within public health, governments invest in groundbreaking initiatives to address the epidemic of mental health disorders and non-communicable diseases. Through innovative programmes focused on prevention, early intervention, and holistic care, they strive to reduce rates of depression, anxiety, and chronic illness, empowering individuals to lead happier, healthier lives.

However, the most inspiring aspect of this future is the synergy that emerges when these two agendas converge. As we heal our planet, we also find ourselves healing – physically, mentally, and spiritually. The air we breathe is cleaner, the water we drink is purer, and our natural spaces are thriving with life. And in turn, our minds are clearer our bodies stronger, and our spirits more resilient.

In the future, every action taken to protect the environment is also an investment in our wellbeing. Whether planting trees to combat deforestation, promoting active transportation to reduce carbon emissions, or creating green spaces to promote mental health, each endeavour is a testament to the transformative power of collective action. Eco anxiety is no longer the main topic of discussion in relation to the psychological impact of environmental actions. Instead, we focus on positive psychological outcomes that make our lives worth living.

So, let us embrace this vision of a future where our planet's and people's health are intertwined in symbiosis and mutual flourishing. Let us work together, across borders and boundaries, to create a world where every individual can thrive, and the legacy we leave for future generations is one of hope, resilience, and abundance. In the convergence of environmental stewardship and human wellbeing is a promise of a future where the earth is healed, and all who inhabit it are alive.

Pro-environmental interventions combined with positive psychology interventions

Let us envision a future where environmental and wellbeing education thrives in schools worldwide. Picture a world where young minds, eager and passionate, take centre stage in leading pro-environmental interventions as they become an integral component of their school's wellbeing program.

In this future schools become not just centres of academic learning but crucibles of transformation. They nurture students' health and wellbeing by instilling in them a deep sense of responsibility towards the planet they call home. From the earliest years of education, children are empowered to become agents of positive change, equipped with the knowledge, skills, and values necessary to tackle pressing environmental challenges head-on.

Imagine classrooms buzzing excitedly as students brainstorm creative solutions to reduce waste, conserve energy, and protect local ecosystems. From organising recycling drives to establishing community gardens, from advocating for sustainable practices within their schools to leading educational campaigns in their neighbourhoods, the possibilities are as boundless as our youth's imaginations. Most importantly, students are not overwhelemed by the task but excited to contribute.

But the impact extends far beyond the confines of school walls. As students engage in hands-on, experiential learning experiences, they become ambassadors for environmental conservation within their families, communities, and beyond. Their harmonious passion and dedication ripple outward, inspiring others to join them in their mission to create a more sustainable and resilient world for future generations.

Moreover, the benefits of integrating pro-environmental interventions into school wellbeing programmes begin to show. Students develop a deeper connection to nature and an improved sense of purpose and fulfilment and experience improvements in their mental and physical health. Engaging in environmental action fosters a sense of empowerment, agency, and belonging, helping them manage feelings of anxiety, depression, and isolation that are all too common among young people today.

In this future, the wellbeing of students and the planet's health are intertwined, reinforcing each other's missions. Through their actions, young people become catalysts for change, proving that when given the opportunity and support, they are capable of shaping a future that is not only sustainable but also filled with hope, possibility, and promise.

So, let us dare to dream of a future where every school has stories to tell about their students wellbeing positively impacted by environmental leadership. Together, let us inspire a generation of eco-literate, empowered, and compassionate individuals committed to building a world where people and the planet thrive in harmony. A generation with a future that is as bright and beautiful as their imaginations can conceive.

Balancing self-care and passion

In our lives, there are moments when we find ourselves at a crossroads – moments when the path we've been following no longer feels right, when the weight of our actions begins to bear us down. In these moments, we are faced with a choice to stay stuck in obsessive passion or embrace a path of harmonious passion guided by informed decisions and effective coping behaviours.

Imagine a world where individuals possess the wisdom and self-awareness to recognise when their pro-environmental actions are causing them distress – when the enthusiasm of their dedication threatens to consume them, leaving them depleted and overwhelmed. In this world, they do not shy away from acknowledging their struggles, nor do they allow themselves to be shackled by guilt perfection or overthinking. Instead, they summon the courage to take a step back, to reassess their priorities, and to recalibrate their approach with compassion and clarity.

In this world, individuals understand that the intensity of one's actions does not measure true passion but by the harmony and balance with which they are pursued. They recognise that sustainable change cannot be built on a foundation of obsession and burnout but on a commitment fuelled by purpose, resilience, and self-care.

And so, they embrace a path of harmonious passion that honours their values, nurtures their wellbeing, and fosters sustainable progress. They engage in effective coping behaviours, seeking out support, seeking out support, practising mindfulness, and finding solace in nature's embrace. They understand that optimism, far from being naive or unrealistic, is a powerful force for resilience and perseverance that helps them get through the darkest of times and propels them forward, even when the road ahead seems daunting.

In this world, optimism trumps pessimism – not because they are blind to the challenges ahead, but because they refuse to be defined and weighed down by the climate change and other environmental issues. They draw strength from the knowledge that every small action, has the power to ripple outward, creating waves of positive change that can change the world.

So, let us dare envision a world where passion is tempered by wisdom, optimism reigns, and individuals dare to recognise when their actions are causing distress. Let us cultivate a world where effective coping behaviours are valued and embraced, and each environmental step forward is guided by compassion, purpose, and resilience. In this world, we have the power to keep going, growing, and making consistent positive change – now and for generations to come.

Final word for now

As we come to the close of this book, we are filled with hope and gratitude. Hope for the possibilities ahead – from the transformative power of positive psychology and the unwavering commitment to environmental stewardship. And gratitude for the journey we've shared – a journey that has challenged us to think differently, to dream bigger, and to believe in the profound connection between human wellbeing and the health of our planet.

Through the pages of this book, we have explored the intersection of two realms – positive psychology and environmental protection – and discovered the possibilities that emerge when these worlds collide. We have seen how the principles of positive psychology can serve as a guiding posts, revealing the path towards sustainable action and personal flourishing. And we have witnessed the profound positive impact that even the most minor acts of environmental stewardship can have on our psychological, emotional, and social wellbeing.

As we reflect on the insights in this book, we are reminded that positive psychology is not just a theoretical framework – it is a foundation for a call to action. It is a call to harness the power of optimism, resilience, and gratitude to serve a cause greater than ourselves. It is a call to embrace a mindset of abundance, recognising the boundless potential within each of us to make a positive difference in the world.

For those already engaged in environmental protection, may this book serve as a source of inspiration and empowerment – a reminder that your work is about saving the planet and nourishing your soul. May you find solace in knowing that every tree planted, every river cleaned, and every species protected is a testament to your commitment to the earth and your wellbeing.

And this book may be a sliver of hope for those just beginning their journey and it may fill their heart with courage and determination. May you take the lessons learned within these pages and allow them to inform and enrich your pro-environmental actions, knowing that in doing so, you are not only protecting the planet but nurturing your growth and fulfilment as well.

Let us carry with us the spirit of possibility, resilience, and unity. Together, guided by the principles of positive psychology, we have the power to create a world where environmental protection and personal flourishing are not just compatible goals but inseparable partners in the journey towards a brighter, healthier and more sustainable future.

Index

Note: **Bold** page numbers refer to tables.

ABCD tool for optimistic thinking 113–114
absorption 51, 106, 165
accomplishment 70, 180, 197; PERMA model 72; personal 34, 72; sense of 107–109
Active Constructive Response (ACR) 162–163
Active Destructive Response (ADR) 163
Adelaide, Australia 45, 49
agriculture: intensification of 5; and surface water bodies 4; sustainable 181, 198; sustainable farming practices 11
air pollution 53–54; and mental health 46–48, 53
altruism 17, 25; competitive altruism 94; and environmental action 94–95; kin altruism 94; mutualism 94, 95; reciprocal altruism 94, 95
American Psychological Association 156
anticipation 109–110
Ash Dieback disease 105, 170
Attention Restoration Theory (ART) 52
Authentic Happiness model 70
Autism Spectrum Disorder 65
autonomy 17, 28, 31, 131, 172, 177–178
awe 97–98

Bamberg, S. 7–8
Barragan-Jason, G. 32
Baumeister, R. F. 185
Bee Well project 75–78, 153; human relationship with nature 76–77; positive health 77–78; wellbeing literacy 76–77
behavioural expression 165
benevolence 96; environmental 7; human 26; and kin altruism 94
Best Possible Eco-Self 143
Bhutan 24
biodiversity 3, 69, 128, 134, 136, 141, 150–151, 153, 170, 172
bio-inspiration/biomimicry 12
biophilia 52
Bo Ket tree 116
Boyd, J. 140
British Columbia 9
Broaden-and-Build theory 83–85, 99, 184
Brown, K. W. 27
Bryant, F. B. 164–165
Buijs, A. 17–18
Burke, J. 75, 77–78
Burton, C. M. 89

capitalisation 162–164; benefits 162; family-based environmental 164; response 162
Capstick, S. 34–35
Carr, A. 114
character strengths 156–162, **156**, 195–197; and pro-environmental actions 157–162, **158–160**; strengths-pro-environmental family contract 160; taxonomy of 156–157; VIA 195–196
Chowdhury, R. 111
citizenship 161–162, 188–189
Clerkin, Sean 13
climate crisis 69
Coexistence Management 9, 11–12
Coghlan, A. 160

collective culture 33–35
collective resilience 88
collectivism 33
community 197–198; autonomous pro-environmental actions 177–178; ecosystem model 171–183; and environmental practice 171; forgiveness 190–191; health and wellbeing 170–192; positivity resonance 183–185; pro-environmental actions 171–183; sense of belonging 185–189; vitality 24
community-based actions 13–15
community-based citizen science projects 182
community-based green businesses 182
community-based initiatives 183
comparing, and savouring 165
compassion 190, 196, 200–201; and Loving-Kindness Meditation 86; nurturing for nature 86; self-compassion 63
competence 25, 28, 73
competitive altruism 94
compulsion, as characteristic of obsessive passion 131
conflict 131–132, 152, 175, 190–191
conscious mindsets 128
consciousness writing 134
Conservation of Resources (COR) theory 187
Contemplative Landscape Model (CLM) 55–56
Convivial Conservation 9
Corral-Verdugo, V. 106, 173
Corrigan, S. 75
counting blessings 165
COVID-19 pandemic 38
Cox, Joe 15–16
creative expression and eco-mood repair 87
creative writing 133
crop cultivation 3
crowding out concept 141–142
cultivating positivity for pro-environmental resilience 87
culture: collective 33–35; individual 33–35
curiosity 65, 106, 161, 170, 180, 196

Dalai Lama 94
Dalebroux, A. 87
Davidson, R. J. 88
Deci, E. L. 69
Diagnostic and Statistical Manual of Mental Disorders (DSM) 156
Diener, E. 71, 73, 82
Disabato, D. J. 69
disparities, and PEB 36
Duality of Passion Model 130, 132

eco-anxiety: and positive emotions 89–98; and wellbeing 68–69
eco-conscious: behaviours 22, 30, 137, 155; choices 161, 165; lifestyles 96, 173
eco-friendly initiatives 182
eco-guilt 69
ecological degradation 69
Eco-Loving-Kindness Meditation 86
eco-mood repair through creative expression 87
ecosystem model 171–183
Emmons, R. A. 92–93
emotional contagion 183–185
emotional wellbeing 67, 72, 82–83; positive emotions 83
emotional wellness 105
employment status 36
engagement 70, 73–74, 106–107; and positive ecology 18
enjoyment 131
environment: in context of illbeing 65–78; in context of wellbeing 65–78; and mental health 66–74; and positive emotions 85–87; vitality 24–25
environmental actions: acts of kindness and 94–95; altruism and 94–95
environmental activism 88, 107, 129–132, 135, 151, 180, 184
environmental catastrophes 48
environmental degradation 68; and climate change 5; and human activity 3–5; overexploitation of natural resources 5; and pollution 5
environmental design: and environmental outcomes 53; and mental health 53
environmental enthusiasts 129, 146
environmental expressive writing 133
environmentalism 105
environmental mastery 172, 178–179
environmental obituary 120–121
Environmental Protection Agency (EPA) 4
environmental psychology 22, 68
environmental sciences 76

environmental stewardship 99–100, 129, 135, 142–143, 150–151, 153–154, 186, 195–201; collective consciousness of 173; collective efforts in 155, 172, 173, 182; and community wellbeing 180; families role in 154; growth mindset in 135; harmonious passion for 130–131; holistic ethos of 174; obsessive passion for 130–132; and psychological wellbeing 171; realm of 135; sense of responsibility towards 183
environmental wellbeing 96–97
eudaimonic (virtuous) values 30
eudaimonic (meaning-based) wellbeing 30, 37, 69
Europe, flourishing in 74–75
European Green Deal 23
European Union (EU): Nature Restoration Law 23; Water Framework Directive 5
evidence-informed model 151
expressive writing 96–97, 133–134
external rewards 37, 132
extrinsic motivations 26–29

family: capitalisation 162–164; character strengths 156–162; children's impact on pro-environmental behaviours of 154–156; congratulations 165; health and wellbeing 150–167; parent-child relationship 152; pro-environmental activities 152–153; relationship 152; savouring 164–166
family-based environmental capitalisation 164
Family Socialisation Theory 154
femininity and PEBs 35
financial disincentives, and PEB 31
fixed mindset 135–137
flexibility 83–84, 89, 131
flourishing 67; in Europe 74–75; model 71, 73–74; psychological richness 75; wellbeing and pro-environmental behaviour 75
"flow state" 51
forgiveness 190–191
Frankl, Victor 121–122
Frederickson, Barbara 83, 87
Fredrickson, B. L. 184
future-focused environmental journaling 110
future-oriented individuals 141

Gable, Shelley 162–163
gender dynamics: and femininity 35; and masculinity 35, 37; and PEBs 35, 37
Genuine Progress Indicator 24
Gifford, R. 173
Gift of Time 96
giving, in environmental context 73
global wellbeing 24–25
Gordon, C. L. 166
gratitude 71, 74, 84, 87; as driver of private/public pro-environmental behaviour 91; for Earth's abundance and biodiversity 165; and interpersonal relationships 91; positive emotions 90–94; practical implementation of 92–94; state 90; trait 90, 92
Gratitude List activity 92–93
"green field sites" 57
green kindness 95–96
green spaces and mental health 51–52
greenwashing 7
Gross Domestic Product (GDP) 24, 38
Gross National Happiness (GNH) 24
growth mindsets 129, 135–137

habitat destruction 5
harmonious passion 130–131
Haslam, C. 187
health 1; community 170–192; and environmental agendas 198–199; environmental impacts to 6; family 150–167; positive 77–78; *see also* mental health; physical health
Healthy Parks Healthy People framework 49
hedonic happiness 166
hedonic wellbeing 30, 69
"helper's high" 26
Hippocrates 50
Hobfoll, S. E. 187
hope 50, 71, 84, 142, 161, 161, 170, 185, 190–191, 197, 199–201; and optimism 111–119; profiling 116–118
hopefulness 115–119
Howell, R. A. 174
human: benevolence 26; relationship with nature 76–77; vitality 17
human activity: and environmental degradation 3–5; and pollution 46
human health *see* health

Huppert, F. A. 74
hydrocarbon leaks 48

illbeing: environment in context of 65–78; introduction to 68
income level, and PEB 35–36
Indigenous peoples 65
individual culture 33–35
individual emotional wellbeing 81–100; Broaden-and-Build theory 83–85; eco-anxiety and positive emotions 89–98; emotional wellbeing 82–83; life satisfaction 98–99; positive emotions and environment 85–87; resilience 87–89
individuals psychological wellbeing 104–124; anticipation 109–110; engagement 106–107; hope and optimism 111–119; purpose and meaning-making 119–124; sense of accomplishment 107–109
industrialisation 4, 45, 58
Integrated Model of Group Membership and Social Identity (IMGMSI) 187–188
International Union for the Conservation of Nature (IUCN) 13
interpersonal relationships 91
intrinsic motivations 26–29
Ireland 4, 13–14, 57
Isen, A. M. 25

Jacobs, M. 17–18
Jetten, J. 187
judgement 161–162, 173

Kahneman, D. 98
Kaplan, Stephen 50
Kasser, T. 27
Keyes, C. 71, 72, 171
Khatri, V. 166
killjoy thinking (reverse) 165
kin altruism 94
kindness: altruism and environmental action 94–95; awe 97–98; benevolence 96; and environmental actions 95; gift of time for greener world 96; green 95–96; impactful contributions 95; positive emotions 94–98; purposeful purchases 95; social recycling 95
King, L. A. 89
knowledge, and PEBs 37
Kuhbandner, C. 83
Kunming-Montreal Global Biodiversity Framework 23

Lagos, Nigeria 55
land-use alterations 3
Leary, M. R. 185
Lenger, K. A. 166
Let It Bee project 13–15, 67, 151, 171
life: purpose 121; satisfaction 98–99; transitions 152
Lithuania 32–33
Liu, Y. 106
livestock farming 3
The Living Standards Framework 24
logotherapy 121–124
Lomas T. 76
Loving-Kindness Meditation 86

MacNamara, A. 166
"Making Hope Happen" programme 116
marine litter 66
masculinity and PEBs 35, 37
Mauss, I. B. 82
McCullough, M. E. 92–93
Meadows, Donella 10; *Thinking in Systems: A Primer* 10
meaning 70, 73, 74; -making 119–124; and positive ecology 18
memory building 165
mental health: and air pollution 46–48; and environmental design 53; and environment connection 66–74; and green spaces 51–52; natural *vs.* urban environments 50–51; and pollution 46–50, **47**; urban design for 53–56, **54**
Mental Health Continuum (MHC) model 67, 71, 72–73, 75; emotional wellbeing 72; psychological wellbeing 72; social wellbeing 72–73
mindfulness 52, 55, 63
mindsets 128–146; conscious 128; fixed 135–137; growth 129, 135–137; maximising vs satisficing 144–145; passion 130–134; pro-enviro behaviour 129, 140–143; shifting/satisficing for environmental impact 145; stress 129, 137–140; time perspectives 140–143
Model of Human Caring (MHC) 50
monetary incentives, and PEB 31

moral obligation 34, 38
Möser, G. 7–8
motivations: behind pro-environmental behaviour 25–26; intrinsic *vs.* extrinsic 26–29
mutualism 94, 95

Natural Capital 4
natural environments: and mental health 50–51; *vs.* urban environments 50–51
nature: human relationship with 76–77; mindfulness in 52; nurturing compassion for 86; role in shaping wellbeing 52–56; therapy 52
nature-based solutions (NbS) 12–16
nature-based stewardship 25
Nature-based Thinking (NbT) 9, 12–16; community-based actions 13–15; role of communication 15–16
Neff, K. D. 173
negative past perspective 140, 142
Nilsson, A. 173

Oades, L. G. 76
Obama, Barack 115
observed variance 8
obsessive passion 130–132
ocean biodiversity 66
Oishi, S. 75
opposing outcomes 132
optimism 73, 74; and hope 111–119; and pessimism 200
orcas 9
organophosphatepesticides 48
Otto, S. 155

parent-child relationship 152
Paris Agreement 23
passion 130–134; for environmental activism 129; harmonious 130; obsessive 130; and self-care 200–201; *see also* Duality of Passion Model
Passive Constructive Response (PCR) 163
Passive Destructive Response (PDR) 163
Passmore, H.-A. 69
Pensini, P. 155
PERMA model 67, 70, 71–72, 75
personal growth 172, 180–183; collaboration and collective action 181; community-driven 181–182; and empowerment 181; and introspection 180; and self-awareness 180; and self-efficacy 181
Personal Growth Initiatives 181–182
Peterson, Christopher 156
physical health 45, 120; expressive writing 133; and pollution 46, 48; and urban design 53
Pihkala, P. 68
pollution: defined as 46; and environmental degradation 5; and human activity 46; and industrialisation 45; and mental health 46–50, **47**; and physical health 46; and urbanisation 45; water 5, 46
positive community psychology 182–183
positive ecology 10, 17–18
Positive Ecology model 70
positive emotions 17, 70, 74, 83, 184; Broaden-and-Build theory 83–85; and eco-anxiety 89–98; and environment 85–87; gratitude 90–94; kindness 94–98
positive environmental experiences 164
positive experiences 89, 164–165
positive health 77–78
positive outcomes 131
positive past perspective 140–141, 142
positive psychology 16–18, 67; positive ecology 17–18
positive psychology intervention (PPI) 16, 76, 91, 141, 199–200
positive relationships 172, 175–177
positivity: and pro-environmental resilience 87; resonance 183–185
present fatalism 141
present-fatalistic perspective 141–143
present-hedonistic perspective 141, 142
pro-environmental behaviour (PEB) 1, 4–5, 106, 128–129, 132–133, 137, 140–143, 185; and character strengths 157–162, **158–160**; children's impact on family 154–156; and collective culture 33–35; collective endeavour 7; community 171–183; community-based Personal Growth Initiatives 182; complexities of 133; defining 23–25; determinants 32–33; environmental challenges 7; and family 152–153; future of 195–201; importance of 150–151; and individual culture 33–35; and individual wellbeing 22–39;

intrinsic *vs.* extrinsic motivations 26–29; motivations behind 25–26; and positive community psychology 182–183; role of 7–9; and self-acceptance 172; shifting/satisficing for environmental impact 145; and social intelligence 161; social norms and values 29–31; socio-demographics of 35–36; strengths and values 157, **158–160**; substantial commitment 172; types of 32; and wellbeing 75; Wellbeing Ecosystem Model 172
pro-environmental interventions 199–200
pro-environmental resilience 87
psychological richness 75
psychological wellbeing (PWB) 67, 72, 105, 133, 171, 172, 175–176, 180, 187; individuals 104–124; model 70
psychology 105; environmental 22, 68; positive 16–18, 67; positive community 182–183
public health 74, 171, 198; advocates 45; policies 58; and wellbeing 74
Public Health Model 74
purpose 119–124; in life 172, 179–180
purposeful purchases 95

Quang Nam Elephant Reserve 115

Reagan, Ronald 114–115
reciprocal altruism 94, 95
relatedness 28, 73
resilience 74, 87–89; collective 88; defined 87; individual emotional wellbeing 87–89; pro-environmental 87; team 88
Riis, J. 98
Rowe, G. 83
Rudd, M. 98
Ryan, R. M. 69
Ryff, C. 105, 171, 172, 175

Samios, C. 166
savouring 164–166
Schultz, P. W. 154
self-acceptance 73, 172–175
self-care 50, 130–131, 200–201
Self-Determination Theory (SDT) 28–29; autonomy 28; competence 28; relatedness 28
self-directed learning 120
self-enhancement values 33
self-esteem 74
self-regulation 157, 161
Seligman, Martin 16, 51, 70, 111, 156
sense: of accomplishment 107–109; of belonging 185–189
sensory perceptual sharpening 165
Sharot, T. 111
Singapore 55–56
Singer, B. 172
Sinkevičius, Virginijus 14
SO, T. T. C. 74
social capital 37
social categorisation 186
Social Identity Theory (SIT) 186–187
social intelligence 161, 162
social norms 29–31
social recycling 95
social wellbeing 67, 72–73
socio-demographics: of PEBs 35–36; of wellbeing 35–36
socio-economic status: educational attainment 35; income level 35–36; and PEB 34–35; and wellbeing 34–35
Socratic Questioning 114
soil erosion 5
Spera, S. P. 134
stakeholder engagement 57
state gratitude 90
Steger, M. F. 69
strengths-pro-environmental family contract 160
stress-is-debilitating mindset 139
stress-is-enhancing mindset 138–140
stress-is-helpful mindset 138, 139–140
stress mindsets 129, 137–140
Stress Reduction Theory (SRT) 52
subjective wellbeing (SWB) 67, 82, 105
surface water bodies 4; and agriculture 4; catchment activities 4; and tourism 4
surf therapy 65
sustainable farming practices 12
"systems thinking" approach 10

Tajfel, Henri 186
Taufik, D. 26
team resilience 88
temporal awareness 165
Thinking in Systems: A Primer (Meadows) 10
Third Culture Kids (TCKs) 188–189

Thunberg, Greta 22
time perspectives 140–143; changing 143; future-oriented perspective 141, 143; negative past perspective 140, 142; positive past perspective 140–141, 142; present-fatalistic perspective 141–143; present-hedonistic perspective 141, 142; for pro-environmental actions 142–143
tourism: Ireland 13–14; and surface water bodies 4
trait gratitude 90, 92
Turner, John 186
2030 Agenda for Sustainable Development 12

United Nations 8; Human Development Index (HDI) 24; Sustainable Development Goals (SDGs) 25
urban design: Lagos, Nigeria 55; for mental health 53–56, **54**; Singapore 55–56
urban environments: and mental health 50–51; *vs.* natural environments 50–51
urban greening 49
urbanisation 1, 4, 45, 56
urban spaces 45, 55–58

Valkengoed, A. M. van 33
values: eudaimonic (virtuous) 30; pro-environmental behaviour 29–31; self-enhancement 33; wellbeing 29–31
Values In Action (VIA) 195–196
Valuing Natural Capital in Communities for Health (VNiC-Health) 57
Veroff, J. 164–165
vitality 74; community 24; environmental 24–25; human 17; physical 77

"warm glow" theory 25
Warren, C. 160
water pollution 5, 46
Wei, X. 75
wellbeing 1; community 170–192; and eco-anxiety 68–69; emotional 82–83; and environmental actions 9, 36–38; environmental impacts to 6; environment in context of 65–78; family 150–167; flourishing model 73–74; global 24–25; individuals' psychological 104–124; introduction to 69–74; mental health continuum 72–73; PERMA model 71–72; and pro-environmental behaviour 22–39, 75; psychological 133, 171, 172, 175–176, 180, 187; role of nature in shaping 52–56; social norms 29–31; socio-demographics of 35–36; and socio-economic status 34–35; values 29–31
Wellbeing Ecosystem Model 171–172
wellbeing literacy 76–77
The Wellbeing of the Nation 24
Westgate, E. C. 75
whole systems approach 9–11
wildfires 48
Wilson, K. A. 166
World Health Organization (WHO) 4, 195
World War II 121
Worthington Jr, E. L. 190
writing: consciousness 134; creative 133; expressive 96–97, 133–134

Zimbardo, P. 140

For Product Safety Concerns and Information please contact our EU representative GPSR@taylorandfrancis.com
Taylor & Francis Verlag GmbH, Kaufingerstraße 24, 80331 München, Germany

www.ingramcontent.com/pod-product-compliance
Lightning Source LLC
LaVergne TN
LVHW010900110826
845149LV00005B/1428

* 9 7 8 1 0 3 2 5 9 0 4 0 0 *